ROYAL COMMISSION
ON
ENVIRONMENTAL POLLUTION

CHAIRMAN: SIR BRIAN FLOWERS

SIXTH REPORT

NUCLEAR POWER
AND THE ENVIRONMENT

Presented to Parliament by Command of Her Majesty
September 1976

LONDON
HER MAJESTY'S STATIONERY OFFICE
£2·65 net

Cmnd. 6618

[redacted]

Previous Reports

1st report	First Report	Cmnd. 4585	February 1971
2nd report	Three Issues in Industrial Pollution	Cmnd. 4894	March 1972
3rd report	Pollution in some British Estuaries and Coastal Waters	Cmnd. 5054	September 1972
4th report	Pollution Control: Progress and Problems	Cmnd. 5780	December 1974
5th report	Air Pollution Control: an Integrated Approach	Cmnd. 6371	January 1976

ISBN 0 10 166180 0

ROYAL COMMISSION
ON
ENVIRONMENTAL POLLUTION

SIXTH REPORT

To the Queen's Most Excellent Majesty

MAY IT PLEASE YOUR MAJESTY

We, the undersigned Commissioners, having been appointed "to advise on matters, both national and international, concerning the pollution of the environment; on the adequacy of research in this field; and the future possibilities of danger to the environment";

And to enquire into any such matters referred to us by one of Your Majesty's Secretaries of State or by one of Your Majesty's Ministers, or any other such matters on which we ourselves shall deem it expedient to advise:

HUMBLY SUBMIT TO YOUR MAJESTY THE FOLLOWING REPORT.

CONTENTS

	Para.	Page
FOREWORD	1	1

CHAPTER I

Introduction

	Para.	Page
The choice of the study	5	2
Arrangement of the report	14	5
Organisational arrangements in Britain	19	6

CHAPTER II

Radioactivity and radiobiology

	Para.	Page
Introduction	25	10
The nature of radioactivity	26	10
The properties of radiation	36	13
Units of radiation	40	15
Current levels of radiation in the UK	43	16
Biological effects of radiation	48	18
Long-term effects	50	18
Genetic effects	57	21
Acute effects	63	23
The biological effects of plutonium	66	23
Movement of radioisotopes in the environment	78	27
Effects on the natural environment	81	28

CHAPTER III

Nuclear power

	Para.	Page
Introduction	82	30
Nuclear physics	83	30
Fissile materials	86	31
The chain reaction	88	32
Critical assembly	90	33
Nuclear reactors	96	34
Types of thermal reactor	102	37
The breeding principle	108	41
The Liquid Metal-cooled Fast Breeder Reactor	112	44
The nuclear fuel cycle	118	47
Uranium mining	119	48
Uranium conversion and fabrication	122	49
Fuel loading and discharge	127	51
Fuel reprocessing	130	56
Radioactive wastes	136	60
Plutonium as a reactor fuel	143	66
Thermonuclear fusion	145	66
Review	153	70

	Para.	Page

CHAPTER IV

Major issues raised by nuclear power

Introduction	154	71
World energy demand	155	71
Projected nuclear development	157	72
Nuclear power and other technologies	162	73
The proliferation of nuclear weapons capability	165	74
Reactor safety	168	76
Radioactive waste	178	80
The diversion of plutonium	182	81
The uncertainty of radiological standards	187	82
The case for nuclear power	189	83
Conclusion	196	85

CHAPTER V

International and national control arrangements

Introduction	198	86
Basic standards	200	86
Radiation standards in the UK	213	90
Derived working limits and discharge limits	215	90
Emergency reference levels	218	91
Arrangements for radiological protection in the UK ..	219	92
The endorsement of basic standards	220	92
Research	227	95
The control of discharges	239	98
Monitoring and surveys..	248	101
Accidental releases of radioactivity	254	102
Conclusion	258	104

CHAPTER VI

Reactor safety and siting

Introduction	260	105
The effects of radioactivity releases	262	105
Safety considerations	269	107
The principles of safety analysis	272	108
Reactor design and construction in the UK	277	112
The Nuclear Installations Inspectorate (NII)	280	112
The maximum credible accident	283	113
Method of working of the NII	287	114
The provision of advice to Government on nuclear safety ..	291	115
Siting policy for reactors	293	116
Fast reactor safety	297	117

	Para.	Page
CHAPTER VII		
Security and the safeguarding of plutonium		
Introduction	305	121
The threat of sabotage	309	122
The diversion of fissile materials	316	124
International safeguards	326	127
Effects on civil liberties	330	128
Security arrangements in the UK	334	129
CHAPTER VIII		
Radioactive waste management		
Introduction	337	131
Gaseous waste	339	131
Liquid waste	348	134
Solid waste (low level)	355	136
Solid waste (intermediate level)	361	138
CEGB and SSEB sites	362	138
AEA and BNFL sites	365	139
Ocean dumping	369	140
High level waste: interim management	373	142
Waste formation	375	143
Vitrification of wastes	381	145
Actinide separation	384	146
High level waste: ultimate disposal		
The range of options	389	148
Disposal to geological formations on land	398	152
Disposal into the ocean bed	409	155
Other issues in radioactive waste management	419	160
Transport	420	160
Foreign reprocessing work	422	161
The costs of sound radioactive waste management ..	425	161
The future: policy and its execution	427	162
CHAPTER IX		
Energy strategy and the environment		
Introduction	436	165
Technical considerations		
Units	437	165
Efficiency and conversion	440	166
Consumer requirements	443	167
Energy supply and pollution	448	168
World energy demand and supply	450	170
UK energy use, past and present	460	174
Official energy strategy for the UK	466	178
Discussion on official energy strategy	474	181
An alternative energy strategy	485	185
Discussion on alternative strategy	492	189

CHAPTER X

Nuclear power and public policy

	Para.	Page
Introduction	498	191
The hazards of nuclear power development	502	192
Aspects of future energy supply	509	194
FBR development	516	197
Public assessment of nuclear power	521	198

CHAPTER XI

	Para.	Page
Summary of principal conclusions and recommendations ..	525	200
Acknowledgements	536	205

FIGURES

		Page
Figure 1.	Map of the UK showing the licensed nuclear sites and AEA establishments	9
Figure 2.	Alpha-activity from decay of 1kg of americium-242m (half-life, 152 yr) and its decay daughters	13
Figure 3.	The curve of binding energy	31
Figure 4.	Schematic diagram of a thermal reactor (Magnox) ..	35
Figure 5.	The changes in numbers of fissile nuclei in Magnox fuel during irradiation, showing the contribution made by Plutonium-239	42
Figure 6.	Schematic diagram of the Prototype Fast Reactor (PFR)	44
Figure 7.	The nuclear fuel cycle	48
Figure 8.	The variation in radioactive decay heating in discharged reactor fuel	54
Figure 9.	The decrease in radioactivity in high-level wastes with time for thermal and fast reactors	62
Figure 10.	Contributions to total alpha-activity at different times in high-level wastes from a commercial fast reactor ..	63
Figure 11.	Sketch of proposed toroidal fusion power reactor (Tokamak)	68
Figure 12.	Diagram showing the organisation of radiological protection in the UK	93
Figure 13.	Incidence of man-made disasters (in the USA)	109
Figure 14.	Incidence of natural disasters (in the USA)	109
Figure 15.	Nuclear reactor accident release frequency limit line ..	110
Figure 16.	Map of the world showing tectonic plates and abyssal hill and swale regions	151

		Page
Figure 17.	Diagram showing the major components of a system for disposing of radioactive waste in the ocean bed ..	157
Figure 18.	Summer and winter demands on CEGB system in 1974/75 including days of maximum and minimum demands ..	169
Figure 19.	Usage of energy and gross national product per head for selected industrial countries in 1972	171
Figure 20.	World energy demand, and a possible means of supply, projected to 2025	173
Figure 21.	Usage of energy per head in the UK since 1900 ..	175
Figure 22.	Energy flow in the UK in 1975	177
Figure 23.	The provision of energy for the UK, projected to 2025, on the "official strategy"	180
Figure 24.	The effect of price changes on sales of gas and electricity to the domestic market in England and Wales since 1953..	183
Figure 25.	The provision of energy for the UK, projected to 2025, on the "alternative strategy"	187

TABLES

Table 1.	The different types of radiation	14
Table 2.	Dose rates in the UK from ionising radiation	16
Table 3.	ICRP recommendations of allowable radiation dose equivalents for radiation workers, per year	21
Table 4.	Radiation doses required to give to various species a lethal dose to 50 per cent (LD 50) within 30 days ..	23
Table 5.	Characteristics of major thermal reactor types	37
Table 6.	The more important isotopes of plutonium, americium and curium	61
Table 7.	Countries operating or expecting to operate commercial nuclear reactors above 100 MWe, with their status under the Nuclear Non-Proliferation Treaty (NPT) ..	75
Table 8.	Probability of death for an individual per year of exposure (orders of magnitude only)	77
Table 9.	Volumes and activities of various types of solid waste accumulated at Windscale in 1974 and estimated for 1985. The comparable figures for high-level liquid waste concentrate are also given	139
Table 10.	Levels of activity permitted by the IAEA in solid waste dumped in the ocean. (Under review)	141
Table 11.	Approximate energy equivalents of 1 TWh	166
Table 12.	Energy supplied to final consumers in the UK, 1975, TWh	176

		Page
Table 13.	The sources of energy used in the electricity supply industry in the UK, 1975, TWh	176
Table 14.	AEA-projected programme of electrical generating capacity installed at end of year indicated, GW ..	179

PLATES

Plate 1.	The Magnox station at Trawsfynydd, North Wales ..	39
Plate 2.	Dounreay Experimental Reactor Establishment, Caithness, Scotland	45
Plate 3.	Uranium enrichment by means of the centrifuge process	50
Plate 4.	Inspection of uranium oxide fuel assemblies	52
Plate 5.	Interior view of the AGR station at Hinkley Point, Somerset	53
Plate 6.	Cooling pond for irradiated fuel elements at the Winfrith 100 MW SGHWR	55
Plate 7.	Transport of irradiated fuel by rail to Windscale, Cumbria	56
Plate 8.	The Windscale works of BNFL on the Cumbrian coast ..	57
Plate 9.	The reprocessing line for Magnox fuel at Windscale ..	58
Plate 10.	Magnox fuel cladding being removed	60
Plate 11.	The stainless steel cooling coils for a new high-level fission product storage tank	65
Plate 12.	The plutonium fuel element factory at Windscale ..	66

APPENDICES

Appendix 1.	Written evidence submitted by the British Institute of Radiology	207
Appendix 2.	Members of the Commission	214
Appendix 3.	Organisations and individuals contributing to the study	216
Appendix 4.	Visits	219

Glossary 220

References 229

Index 232

"If a problem is too difficult to solve, one cannot claim that it is solved by pointing to all the efforts made to solve it."

Hannes Alfvén. *Energy and Environment*
Bulletin of the Atomic Scientists. May, 1972.

FOREWORD

1. As a standing Royal Commission on Environmental Pollution we are able to consider any subject in our field where we believe that there is a need for independent investigation. It is an important part of our remit that we should consider long-term possibilities of danger to the environment, of a kind which may not receive adequate attention from government departments involved with more immediate problems.

2. Over the last few years there have been signs of increasing anxiety in many countries about projected growth in nuclear power and about the environmental risks that this might imply for the future. Early in 1974 we decided that this was a fitting subject for us to investigate and we announced the terms of a study of radiological hazards with particular reference to the development of civil nuclear power.

3. We invited evidence from the many organisations having a specific interest in the subject and we have received much additional evidence from other bodies and from the general public. A list of all those who have contributed is given in Appendix 3. We leave until the end of this Report our general acknowledgement of the help we have received, but we wish here to record a special debt of thanks to the United Kingdom Atomic Energy Authority who have been unfailingly helpful in their response to our many enquiries.

4. Finally, we think that we should here acknowledge a fact that may be regarded by some as having a bearing on the independence of our enquiry (though we confidently expect that on this point our findings will speak for themselves); namely, that the Commission's Chairman, Sir Brian Flowers, is a part-time member of the UKAEA Board. We wish simply to record that no pressure of any kind has been brought to bear on the Chairman by the Authority.

CHAPTER I

INTRODUCTION

The choice of the study

5. There are few subjects in the field of environmental pollution to which people react so emotionally as they do to radioactivity. One reason for this is certainly the association with the destructive uses of nuclear energy. Historically, the awesome power that could be released from the atomic nucleus was first dramatically demonstrated by the bombs that were dropped on Hiroshima and Nagasaki, and which terminated the Second World War. Most people who are alive today have grown up in the nuclear era, in which nuclear weapons have been deployed in increasing numbers by one state after another, and in which awareness of the appalling consequences of their use has itself provided a precarious basis for peace between the powers possessing them. In the early post-war years the materials and the knowledge required to make nuclear weapons were held by very few and were closely and jealously guarded, but the spread of peaceful nuclear technology to many nations is making this knowledge and these materials widely available, so contributing to the proliferation of such weapons and to the risk that they may one day be used.

6. The development of reactors to harness nuclear energy for the generation of electrical power stemmed directly from the weapons programmes. Early power reactors were operating in this country and in the USA by the latter part of the 1950s. By the mid-1970s about 200 commercial reactors were in operation in some twenty countries. On current plans, the number of reactors in operation by the mid-1980s will be about 500 in more than thirty countries. Some projections of longer-term growth(1) suggest that by year 2000 the total installed nuclear generating capacity throughout the world will be some thirty times the present capacity. For the UK, as we discuss in detail in a later chapter, projections of growth in nuclear capacity that have been provided in evidence to us indicate a twenty-fold increase by year 2000 and a further quadrupling by year 2030.

7. The increase that has occurred in nuclear capacity, which has been especially marked during the early 1970s, and the rapid growth that is envisaged for the future, reflect the concern of governments to achieve security in energy supplies in the face of uncertainties about the future availability of fossil fuels. The sudden increase in the price of oil in 1973 provided a sharp warning to industrialised countries of their vulnerability on energy supply, and a powerful stimulus to lessen their dependence on this source. Though there are considerable variations in estimates of reserves of fossil fuels, it is certain that these reserves are not inexhaustible. Estimates of world oil reserves are such that if consumption were to continue to rise at the rate that has prevailed over the last decade, the reserves would last until only about the turn of the century. On the long-term perspective that we are here concerned with the supplies of oil under the seas

around our coasts appear as no more than a temporary alleviation. The world reserves of coal are relatively very large but it is by no means clear that coal could be mined on the scale that would be needed to meet growing energy demands, or that the environmental problems arising from its extraction and use would be acceptable. Moreover, it must be remembered that fossil fuels will be needed not only as a source of energy but as raw materials for the chemical industry. It has been said that our age will be condemned in retrospect for its profligate use of these fuels. Against this background the emergence of nuclear power as an alternative energy source for mankind appears providential.

8. On the other hand, nuclear power introduces environmental risks and problems, and some of these appear unique in their implications for society. There is the problem of dealing with the highly radioactive wastes which arise in the nuclear fuel cycle and which will have to be contained for immense periods of time. There is the problem of the creation of hazardous materials, especially plutonium, which may be used in malevolent acts against society. There is the risk of dangerous releases of radioactivity from reactors or other nuclear installations, whether by accident or sabotage. Much effort has been, and is being, devoted to seeking adequate technical and organisational answers to these problems but it is important that this should not obscure underlying issues which are political, social and ethical in character and which deserve wide public debate. Such debate has so far been most notably and vigorously conducted in the USA, though it is developing in other countries. It is a reflection of the range and difficulty of the issues that the debate appears to emphasise the polarisation of view between those who are for or against nuclear power rather than to give promise of a reconciliation. The proponents of nuclear power see its development as inevitable if the world's energy needs are to be met, and believe that the admitted hazards of that development can be reduced by technical and other safeguards to the level where they should rightly be accepted in relation to the benefits. The opponents of nuclear development see it as a step fraught with future danger for mankind and as one which could and should be avoided by seeking alternative and less hazardous sources of energy, or even by accepting restraints in energy use which might imply significant changes in attitude towards economic growth.

9. Our broad remit is to advise on the future possibilities of danger to the environment. Early in 1974, and in the light of preliminary consultation with those bodies principally concerned with civil nuclear matters in the UK, we decided that the environmental problems that would be associated with a greatly expanded nuclear power programme would be a fitting subject for us to study. The UK, like other industrialised countries, then appeared to stand on the threshold of a major commitment to nuclear power. It was not clear to us that the environmental implications of such a commitment had been sufficiently examined, or that the organisational arrangements and procedures that exist to control nuclear development and to safeguard the public and the environment were adequate for an expanded future programme. Since we started our study the issues raised by nuclear power have been increasingly debated, and this has confirmed our original judgment in undertaking it. We have noticed that the debate is not always well-informed, that sometimes relatively minor matters

receive attention to the exclusion of others potentially more important, and that the context is often poorly defined. It was an important aim of the study that we should present the main facts about radioactivity and nuclear power clearly and simply in our Report, and thus enable the issues to be discussed objectively and in a proper perspective.

10. We realised from the outset that the study would involve us in some matters that were beyond our competence and even our terms of reference. Thus, in attempting to assess the environmental impact of nuclear development we could not ignore the risks and possible consequences of reactor accidents, though an authoritative judgment on this subject would call for expert knowledge of nuclear technology and the techniques of safety analysis which we do not possess. Again, elucidation of the social costs and benefits of nuclear power raises such questions as the nature and implications of the security measures needed to safeguard hazardous materials, and of the extent of future need for this source of energy having regard to possible alternatives and to estimates of energy requirements. It is worth remarking here that the aspects of nuclear development that cause most dissension tend to be those that relate not to normal operation but to the possibility of abnormal events such as accidents at nuclear installations or malevolent acts against them. We have felt bound to consider such issues because of their importance to a total assessment of the potential impact of nuclear power and also because it did not appear to us that they had been expounded, or were likely to be, by any other official body. On some of the more specialised aspects of the study we have consulted independent experts and we have been greatly assisted by their advice. In those areas where we have felt unable to reach a definitive judgment, whether because the factors involved are highly technical or because their nature is such that we have not thought it our proper function to enquire deeply into them, our aim has been to illuminate the issues and to point to those that appear to need consideration by other bodies.

11. We have limited our study of radioactivity very largely to that arising from the nuclear power programme and we have merely touched upon the question of other sources of radioactivity, such as those used in medicine and in industry. The amount of radioactivity involved in these uses is many orders of magnitude* less than that applying to nuclear power (though it should be emphasised that the exposure of members of the public to radiation arising from the routine operation of nuclear power is much less than that from the medical use of radiation in diagnosis and therapy, and from radiation from natural sources). There are possible uses for nuclear power other than electricity supply and we have considered, in particular, whether there were likely to be developments in the use of reactors for merchant ship propulsion such that we should need to assess their environmental effects. In the light of the report by the Nuclear Ship Study Group([2]), however, it appears that there are major uncertainties about whether these ships will be developed on a significant scale. Moreover, many of the problems and risks that arise with marine use are similar in principle to those arising from land-based nuclear plants.

* One order of magnitude is a factor of 10; two orders of magnitude a factor of 100, and so on.

12. We have not considered the military aspects of nuclear power, which are clearly beyond our terms of reference, though we are deeply concerned about the extent to which the proliferation of nuclear weapons may be abetted by the spread of civil nuclear technology. The threat posed by arsenals of nuclear weapons would commonly be regarded as perhaps the gravest facing mankind, and it is one which the world must live with for the foreseeable future. By comparison, and the proliferation problem apart, to question whether the assuredly lesser risks stemming from the beneficent use of nuclear power can safely be entertained might appear to show a want of proportion. But these risks, though less overt, may be more insidious. Nuclear weapons are properly subject to military security, and their control and possible use are certainly matters of very great political consequence. One of the major anxieties that is expressed about the widespread development of nuclear power is that the adequate safeguarding of the increasing quantities of potentially dangerous materials that this will create may lead ultimately to the need for security measures in a civilian context so pervasive as to be damaging and unacceptable in a democratic society.

13. In our study we have naturally been concerned mainly with nuclear power in the UK. As we have noted, however, nuclear development is a world matter and the resulting problems are world problems. The UK cannot be viewed in isolation. Our general terms of reference as a Royal Commission embrace international as well as national considerations and we have therefore not been inhibited from taking the former into account to the extent that appeared necessary in our enquiries.

Arrangement of the report

14. Some of the subject matter of the study is inevitably very technical. We have perforce had to educate ourselves on these matters in order to appreciate the issues and reach conclusions. We hope that our Report will be read by many people who are not familiar with the subject and we have therefore given, where appropriate, an elementary exposition of the technical principles involved. Thus, in the next chapter, in which we discuss radioactivity and its effects on living matter, we have begun with a simplified account of the nature of radioactivity and of the units in which it is measured. We consider in this chapter a question which is fundamental to the study, namely the biological effects of plutonium. In Chapter III, after a simplified account of the principles of nuclear power, we describe the nuclear fuel cycle. Much of this chapter is concerned with the nuclear reactors which produce electricity, but no less important are the other parts of the cycle. The management of the waste products, which is one of the important aspects of the study identified in our original announcement, is briefly described, but we return to this subject in more detail in Chapter VIII.

15. In Chapters II and III we have been largely concerned to provide the factual foundation for the study. On this basis we turn, in Chapter IV, to

consider those issues that are the cause of most anxiety about nuclear development and of opposition to it. To a large extent this chapter constitutes a preliminary discussion of matters which we examine further, and separately, in later chapters. We thought that this discussion was necessary early in our Report, and even at the price of some repetition, so that the issues might be seen as a whole at the start. We are very conscious of the difficulty of these issues, and that their resolution must largely rest on subjective judgments. From this discussion we arrive at some general principles which are relevant to the consideration of more specific issues in later chapters.

16. In Chapter V we describe the national and international organisations that are responsible for the control of radioactivity and nuclear power, and give our conclusions and our recommendations for changes in these arrangements. In Chapter VI we discuss the principles involved in appraising reactor safety and how they should be implemented. Our main aim here has been to consider the nature of the risks and to relate nuclear and non-nuclear hazards. We also discuss siting policy for reactors. Chapter VII is concerned with the security arrangements required in relation to nuclear power. This is clearly a sensitive and specialised subject and it is not one that we have thought it proper for us to consider, or to discuss, in detail. We decided, however, that our study would be incomplete if we were not to seek some understanding of how the security implications of nuclear development are seen by the relevant authorities, and to provide a broad account of our findings. Chapter VIII is about the management of radioactive wastes: what is being done and what we should like to see.

17. We are not concerned, as a Commission, with the formulation of energy policy, though we are certainly concerned with its environmental implications. It appeared to us, however, that some appreciation of current energy consumption in the UK, of future projections of energy demand, of the ways in which this demand might be met and of their environmental consequences, was essential if the development of nuclear power was to be seen in proper perspective. We have considered this subject in Chapter IX.

18. At this stage in the Report we have completed our account of our investigations and findings. In Chapter X we draw these matters together in a discussion which presents our views on the development of nuclear power in this country. We are conscious that in this chapter and elsewhere in our Report we have reached conclusions that will be controversial, and will be challenged. But it is our duty to be a guardian of the future environment and we must report what we believe to be right. Finally, our conclusions and recommendations are summarised in Chapter XI.

Organisational arrangements in Britain

19. Throughout this Report we shall need to refer fairly frequently to the main organisations concerned with radioactivity and nuclear power. We

thought it would be helpful to end this chapter with a brief survey of these organisations, introducing the acronyms by which they are usually known and which we shall subsequently employ. We shall consider relevant aspects of organisation in detail elsewhere in the Report, and especially in Chapter V.

20. After the Second World War, Britain developed nuclear energy for defence purposes under the Ministry of Supply, and during this time the research establishment at Harwell and the production facilities at Capenhurst, Springfields and Windscale were built, and engineering development was undertaken at Risley. The UK Atomic Energy Authority (AEA) was established by Act of Parliament in 1954, to take over both the defence and civil projects of the Ministry of Supply, and with a mandate to "produce, use and dispose of atomic energy and carry out research into any matters connected therewith". The AEA develop prototype power reactors at Dounreay, Windscale and Winfrith and, in conjunction with the design and manufacturing organisation (see below) and the utilities (the Central Electricity Generating Board (CEGB) in England and Wales and the South of Scotland Electricity Board (SSEB)), they carry forward the development to the stage where commercial plants may be ordered.

21. Over the years, the AEA has been the Adam from whose ribs a number of different organisations have been created. From the engineering group at Risley came a number of reactor designers to participate initially in five industrial consortia, later in two—The Nuclear Power Group (TNPG) and British Nuclear Design and Construction (BNDC)—and finally in one, the Nuclear Power Company (NPC), a wholly-owned subsidiary of the National Nuclear Corporation (NNC)*. With the development of commercial nuclear plants, the Nuclear Installations (Licensing and Insurance) Act was passed in 1959: this created the Nuclear Installations Inspectorate (NII). The NII was within a Government Department, latterly the Department of Energy, until 1975, when it transferred to the new Health and Safety Executive (HSE). It oversees the design and planning of plants and processes, and approves their operation, on licensed nuclear sites. The AEA sites do not require this statutory licensing, but their safety arrangements are made in consultation with the NII.

22. In 1971, the AEA responsibilities were further divided. The development of nuclear weapons, primarily carried out at Aldermaston, was made the responsibility of the Ministry of Defence, and henceforward the AEA was concerned only with the civil side of nuclear power. Two commercial undertakings, which had developed within the AEA since 1965, were established as private companies, but all their shares are held by the Authority on behalf of the State. The Radiochemical Centre Ltd (TRC) at Amersham manufactures and sells radioisotopes for use in industry and medicine. British Nuclear Fuels Ltd (BNFL) at Risley operates fuel cycle services—uranium processing at Springfields, uranium enrichment at Capenhurst, fuel reprocessing and

* The state has only a minority shareholding, and the NNC is thus the only privately controlled organisation in the nuclear business in the UK.

plutonium fuel fabrication at Windscale. (See Fig. 1). The AEA remained the research and development organisation with continuing responsibility for technical support to the nuclear industry, and for the development of new reactor types such as the Steam Generating Heavy Water Reactor (SGHWR, see paragraph 106) and the Fast Breeder Reactor (FBR, see paragraph 112). The Authority also remain responsible for the provision of advice to Ministers on all matters pertaining to the civil application of atomic energy.

23. The National Radiological Protection Board (NRPB) was established in 1970 as a result of the Radiological Protection Act 1970. It was required to provide a national, authoritative point of reference on radiological protection, to conduct research and to provide advice in the field. It took over the functions of the Radioactive Substances Advisory Committee appointed under the Radioactive Substances Act 1948 and the Medical Research Council's (MRC) Radiological Protection Service, as well as some activities previously conducted by the AEA. Members and staff of the NRPB contribute to the work of the International Commission on Radiological Protection (ICRP), an unofficial body elected every four years by the International Congress of Radiology, a professional gathering. The ICRP recommend basic standards for radiological protection which are accepted world wide.

24. The protection of the public from radiation is the direct responsibility of Government Departments. The Department of Health and Social Security (DHSS) are concerned with the use of radiation in medicine. Disposals of radioactive waste from licensed sites in England and Wales are the joint responsibility of the Department of the Environment (DOE) and the Ministry of Agriculture, Fisheries and Food (MAFF), each of whom appoints inspectors for the purpose. Those from the DOE form the Radio Chemical Inspectorate (RCI), which is also concerned with the disposal of small quantities of radioactivity onto local tips and into sewers. The control of discharges of radioactivity to atmosphere is the concern of HM Alkali and Clean Air Inspectorate (ACAI), which was formerly part of the DOE but was transferred last year to the HSE. In Scotland the functions of these three Inspectorates are combined in one body, HM Industrial Pollution Inspectorate (IPI).

FIGURE 1

Map of the UK showing the licensed nuclear sites and AEA establishments.

CHAPTER II

RADIOACTIVITY AND RADIOBIOLOGY

Introduction

25. In this chapter we have set down the main features of radioactivity to help in understanding the issues discussed in later chapters. Some technicalities are unavoidable but we have tried to minimise them and to give a simplified description in order to assist the general reader. We have described in some detail the effects that ionising radiation may have on living organisms. A knowledge of these effects is essential to judgement of the balance of advantage and disadvantage that is inherent in the determination of permitted levels for radiation exposure; ignorance of these effects leads to mistrust of the judgements that are made. We have not attempted to give a general account of the characteristics of individual radioactive substances, except for plutonium which is of prime concern because of its great toxicity and the large scale on which it may be produced in the nuclear power programme. For this element we have sought the help of independent consultants to enable us to review its biological effects.

The nature of radioactivity

26. Radioactivity is the emission of certain radiations by the nuclei of unstable atoms. An atom consists of a central nucleus which contains almost all the mass of the atom and which electrically is positively charged, and a surrounding cloud of planetary electrons of very little mass which are negatively charged. Normally an atom is electrically neutral, the central positive charge on the nucleus being exactly balanced by the negative charges on the electrons. The number of electrons in the neutral atom, and hence the charge of the nucleus, determines the element to which the atom belongs and hence its chemical properties. In the course of chemical or physical processes it is common for atoms to gain or lose one or more of their planetary electrons without affecting the charge of the nucleus, and thereby to acquire an overall negative or positive charge respectively: in such a condition the atom is referred to as an "ion" and the process is called "ionisation".

27. The atomic nucleus is a compact blob of extremely dense matter which may be considered to consist of a mixture of two kinds of similar particles—protons and neutrons. A single proton constitutes the nucleus of the hydrogen atom; it has a positive electrical charge equal in magnitude to the negative charge of an electron. It is, however, nearly 2,000 times as heavy. The neutron has very nearly the same mass as the proton but it is electrically neutral. The number of protons in the nucleus thus determines the position of the atom in the periodic table of the elements. Between neutrons and protons in close contact there are very strong forces, the so-called nuclear forces, which are capable of binding them together into a stable nucleus in spite of the electrical

repulsion which exists between the protons. The stability, however, depends upon a rather precise ratio of neutrons to protons. If there are too many or too few neutrons the nucleus will be unstable and will remedy the situation by spontaneously changing the ratio.

28. This it can do in one of two ways, depending upon the precise nucleus concerned. It may emit a tightly-bound chunk of nuclear matter called an α-particle which consists of two neutrons and two protons and is actually the nucleus of a helium atom. In so doing the nucleus loses two positive charges and so transmutes itself into an atom belonging to a chemical element two places lower in the periodic table. This process of spontaneous transmutation accompanied by the emission of α-particles is known as α-radioactivity.

29. Alternatively, the nucleus may undergo a process known as β-radioactivity in which a neutron spontaneously changes into a proton, or vice versa, so that the resulting atom belongs to an element one place higher or lower in the periodic table. In order to conserve electric charge a β-particle is emitted which consists of an electron (if a neutron becomes a proton) or its positive analogue, a positron (if a proton becomes a neutron).

30. The process of going from a less stable to a more stable state releases energy. This energy is used in propelling the α or β-particles which are therefore emitted from the atom with considerable speed (usually a significant fraction of the velocity of light). Being electrically charged they interfere with the electron clouds of atoms through which they pass, thereby causing ionisation and often rupturing the chemical bonds which join atoms together to form molecules. Some of the excess energy may also appear in the form of γ-rays, which are electromagnetic radiations similar to X-rays but usually of shorter wavelengths, and which like X-rays are also ionising.

31. Some very heavy unstable elements which always have a large neutron excess display a third form of instability in which their nuclei break into two large fragments accompanied by a few free neutrons. This is the process of "spontaneous fission" in which the heavy elements decay into two elements in the middle region of the periodic table. The "fission products" are themselves unstable, being usually β-radioactive. The fission process is accompanied by a very large release of energy which mostly manifests itself in the motion of the two fragments. The fragments are stripped of all or most of their electrons, are hence electrically charged, and heavily ionising.

32. Atoms of the same chemical element always have the same number of protons in their nucleus, and hence the same number and configuration of orbiting electrons. But they may have different numbers of neutrons. Such different atoms are called "isotopes"* of the element and have identical chemical properties. They are written e.g. strontium-90 or sometimes strontium^{90},

* The name indicates that they have the same place (in the periodic table).

connoting that the strontium nucleus (which has 38 protons) has 52 neutrons, making a total of 90 "nucleons". Some isotopes are stable, others such as strontium-90 emit radiation and are called "radioisotopes". A substance containing unstable atoms is described as radioactive and, as we have seen, may emit four kinds of radiation: α, β, γ, and if fission occurs, neutrons. The individual processes by which radioactive atoms approach stability by emitting radiation occur completely at random and are independent of all physical and chemical circumstances; the process can neither be advanced nor delayed by man. There is, however, for each particular type of radioisotope, a definite characteristic rate at which the atoms will disintegrate or "decay": this is measured by its physical "half-life", that is, the time in which one half of the atoms will decay. The half-life may be a fraction of a second or it may be millions of years, but it is always the same for a given isotope. After one "half-life" the amount of radioactivity from the given nuclei will be halved, and after a second half-life it will be halved again, and so on. Thus the radioactivity will never reach zero, though after sufficient half-lives it will become negligible.†

33. It follows that for a given number of radioactive atoms the activity (i.e., the rate of radioactive disintegration) is inversely proportional to the half-life. Hence a short-lived substance will be extremely radioactive although for only a short time, while a long-lived substance will be much less radioactive but over a very long period. The unit of measurement of radioactivity is the "curie", abbreviated Ci. One curie corresponds to the amount of activity displayed by one gram of radium (radium–226, whose half-life is 1602 years), namely 37 billion ($3{\cdot}7 \times 10^{10}$) disintegrations per second. After the passage of one half-life, the number of curies associated with the particular radioisotope will be halved. If a radioisotope decays more slowly than radium, that is, it has a longer half-life, then one gram will generally contain only a fraction of a curie. (One thousandth of a curie is a millicurie, mCi, and one millionth, a microcurie μCi.) Short-lived radioisotopes may contain thousands of curies (kCi) or even millions of curies (MCi) per gram: their radioactivity is thus very concentrated. For the purposes of radiation protection it is the radioactivity, rather than the mass, of material that is important. For this reason it is usual to express the amounts of radioactive substances in terms of the curies they contain.

34. Often a radioisotope decays to form a second one, which in turn decays to form a third; the process may continue through many stages, each having its characteristic half-life. The sequence is called a "decay chain", and the subsequent radioisotopes are referred to as "decay daughters" of the first. An example is the decay of uranium–238, the principal constituent of natural uranium, which is found in nature because its half-life (4·5 billion years) is comparable with the age of the earth. It is α-radioactive and decays to thorium–234, which is itself unstable. The chain of "decay daughters" following thorium–234 includes radium–226 (which is found in nature associated with much larger quantities of uranium, as in pitchblende ore) and a number of

† Strictly, radioactivity will reach zero when the last nucleus disintegrates.

very short-lived radioisotopes such as radon and polonium. The final member of the decay chain is lead–206, which is stable.

35. The number of curies of a particular radioisotope in a piece of material will gradually decline as it decays. However since there may be a number of decay daughters each with associated radioactivity, the total activity contained in the material and its associated biological hazard may not decay uniformly with time, but actually increase. For example, Fig. 2 shows the decay over 10^6 years of an isotope of americium. This element does not occur in nature but is created in nuclear reactors by the process described in paragraph 137. During its decay, another isotope of americium, and then ones of curium, plutonium, and uranium are formed in turn. The α-activity increases to a maximum, then declines, and then rises to another maximum as thorium–230 and radium–226 are formed by successive decay.

FIGURE 2

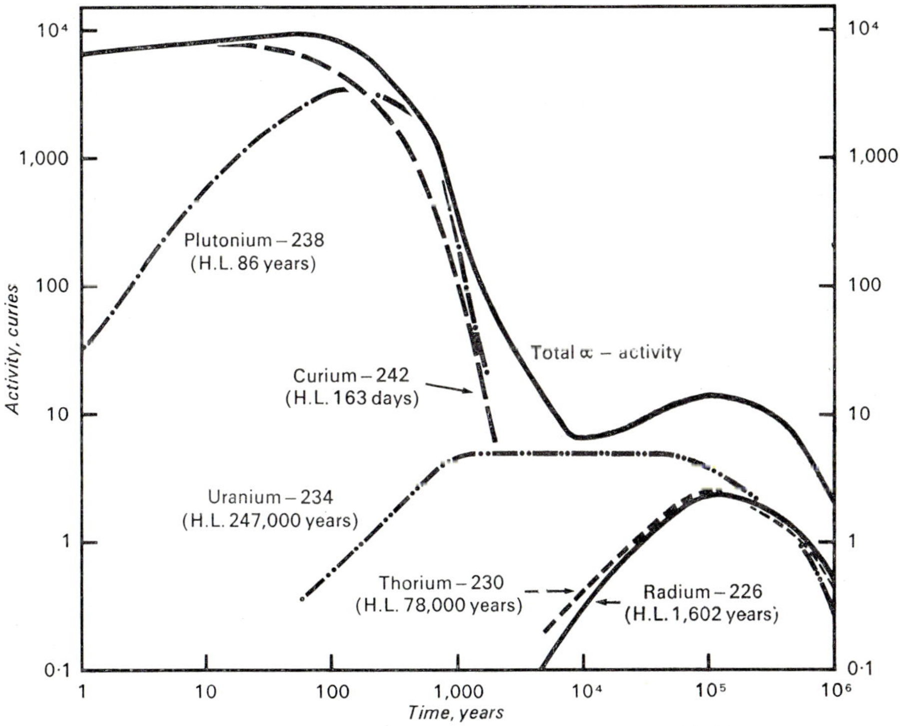

Alpha-activity from decay of 1kg of americium-242m (half-life, 152 yr) and its decay daughters.

The properties of radiation

36. The four kinds of radiation differ in their powers of penetration and in their interactions with the materials through which they pass. They are listed

in Table 1. α-particles are relatively slow-moving and lose their energy in a short distance. They are able to penetrate only a few tens of mm of air and are easily stopped by a sheet of paper. However, because they leave a short but dense trail of ionisation in the matter through which they pass, they can cause more damage in living tissue than particles of longer path length. Nuclei emitting only α-particles, or "α-emitters", are biologically ineffective unless they are taken into the body. This may be by inhalation, ingestion, or at the site of an open wound. But both α- and β-emitters, especially the latter, may also emit γ-radiation.

TABLE 1

The different types of radiation

Type	*Symbol*	*Particle*	*Stopped by*
Alpha	α	Helium-4 nucleus (Heavy, +ve charge)	A few cm of air: 40 μm tissue
Beta	β	Electron (Light, —ve charge)	A few mm of plastic: 40 mm tissue
Gamma	γ	Photon (Electromagnetic radiation)	Progressive attenuation especially by heavy nuclei; e.g. 40 mm of lead reduces to 1/10
Neutrons	n	Neutron (Heavy, uncharged)	Progressive attenuation especially by light nuclei; e.g. 0·25 m of water reduces to 1/10

37. Electrons, or β-particles, vary widely in their energy and it is less easy to specify a range for their effect. Moreover they are easily deflected because their mass is so small. When energetic β-particles are deflected they emit an electro-magnetic radiation called "bremsstrahlung" or "braking radiation" which is like γ-rays and able to penetrate further than the originating electrons. Nevertheless, β-particles lose most of their effect if shielded by say a few mm of perspex. Thus nuclei which are "β-emitters" are again of particular danger if taken into, or onto the surface of, the body. In practice, however, many "β-emitters" also emit very penetrating γ-radiation which, like X-rays, can pass easily through matter, and thus irradiate the whole body. Living cells require to be shielded from γ-radiation by considerable thicknesses of heavy materials such as lead or concrete. Reactors, for example, are usually surrounded by a "biological shield" several metres thick. Gamma-rays are progressively attenuated to any desired level given sufficient thickness of material.

38. Neutrons, because they are electrically uncharged, are slowed down only by direct collisions with nuclei: they pass through atoms almost independently of the electron cloud. After they have lost most of their energy by making many such collisions they will be absorbed into a nucleus, which thus becomes a higher isotope of the same element and will often be radioactive. Absorption of neutrons takes place most efficiently at low energies because it depends (amongst other factors) upon the time the passing neutron spends in the neighbourhood of the capturing nucleus. Since nuclei are very small compared with

atoms, and neutrons uncharged, collisions happen infrequently. Neutrons may therefore penetrate matter to considerable depths before absorption takes place. When a neutron strikes a nucleus, some of its energy will be transferred to the nucleus which therefore recoils. Being electrically charged and massive the recoiling nucleus creates dense ionisation over a short distance like an α-particle does. This is why neutrons, although they do not themselves cause ionisation, are so damaging to living tissue—and, indeed, to solid matter such as fuel elements and their cladding. When neutrons are eventually absorbed, moreover, intense γ-radiation is usually emitted by the absorbing nucleus and this, too, is ionising.

39. Neutrons transfer energy more readily in the slowing down process if the nuclei with which they collide belong to light elements like hydrogen, carbon and oxygen. Slowing down media (or "moderators") typically consist, therefore, of water or graphite, although organic substances can also be used. In shielding against neutrons it is necessary to slow them down with a moderator, to capture them in an absorbing material, and then to shield against the penetrating γ-radiation which follows absorption. All three processes take place if sufficient thicknesses of, say, water are available, because hydrogen absorbs neutrons to some extent, as well as slowing them down, and γ-radiation although capable of penetrating large thicknesses of water is eventually absorbed by it. If economy of space is important, however, the problem of shielding becomes much more complex and multi-layered structures may be necessary.

Units of radiation

40. We have described the nature of ionising radiation and we now turn to its effects on living organisms. It is necessary first to introduce the units used to express the amount of radiation absorbed by a body. When radiation interacts with matter it deposits energy and the amount of this energy in relation to the mass of matter involved is used as a measure of the intensity of radiation. The unit of absorbed dose is the "rad" defined as the quantity of radiation that will cause 1 kg of material to absorb 0·01 joules. This is a very small amount of energy (it is enough to raise the temperature of tissue by only two millionths of a degree) and the damage occurs by virtue of the ionisation produced.

41. The rad is a definite physical quantity which can be measured. Ultimately, however, it is the biological effect of the radiation that is important, and this is much more difficult to measure. For one thing, the different kinds of radiation cause differing amounts of damage to living tissue for the same energy deposited, that is, for the same number of rads. Moreover, these relative effects depend greatly on the nature of the tissue involved and the other conditions pertaining to the irradiation. In principle, in order to measure the biological effect of a particular radiation dose it would be necessary to multiply the amount in rads by a quantity expressing relative biological effectiveness, and which would take account of all these factors. For practical purposes, however, a more approximate approach is taken which allows simply for the effects of the different kinds of radiation. A "quality factor" is defined which is unity for β and γ-radiation and 10 for neutrons and α-particles. This recognises

the fact that for a given number of rads the biological effect of neutrons or α-particles is on average about 10 times as great as that of β or γ-radiation. The product of radiation dose in rads and the quality factor gives the "dose equivalent" in terms of a unit called the "rem". The allowable dose (strictly, the allowable dose equivalent) for a given organ or part of the body is expressed in rems. Thus, an allowable dose of, say, 5 rem could be made up of 5 rads of γ-radiation or 0·5 rads of neutrons. This enables exposure to different types of radiation to be summed at least roughly in terms of their biological effects and compared with prescribed dose limits.

42. For simplicity we shall describe all radiation doses in rems or in the sub-multiple, the millirem (mrem), even though they were given in the original publications in rads. In radiological protection, the doses that will cause particular medical effects are not known with great precision anyway, and discrepancies of even as much as ±30 per cent are usually not very significant for evaluation of the risks.

Current levels of radiation in the UK

43. In this report we are concerned with the radioactivity and radiation doses associated with the nuclear power programme, but it is important to realise that any radiation thereby received would be additional to the radiation that is already received by the general population. Radiation doses from man-made sources are no different in character from naturally-occurring ones, although they may affect particular parts of the body rather than the body as a whole. Table 2 shows the doses, both natural and man-made, currently being received by an average member of the population.

TABLE 2

Dose rates in the UK from ionising radiation([3])

	Bone marrow		*Reproductive cells: genetically significant dose (GSD)*	
	mrem/yr		*mrem/yr*	
Naturally occurring:				
From cosmic rays	33		33	
From soil and airborne	44		44	
Within the body (mainly potassium-40)	24		28	
	—	101	—	105
Manmade:				
Medical, diagnostic X-rays	32		14	
Medical, radiotherapy	12		5	
Medical, radioisotope use	2		0·2	
	—	46	—	19
Fallout from bomb tests	6		4	
Occupational doses (other than from nuclear power)	0·4		0·3	
Nuclear power industry	0·25		0·2	
Miscellaneous (mainly occupational doses)	0·3		0·3	
	—	7	—	5
		154		129

44. The table shows the dose rates for the bone marrow and for the reproductive organs (testes and ovaries, collectively called "gonads"). The reason for selecting these is because of their respective importance in relation to the effects of radiation in producing disease in individuals exposed to it ("somatic" effects) and in its effects on their offspring ("genetic effects"). We discuss these effects further below. Whereas radiation doses to the bone marrow could in principle be harmful to anyone, and therefore the average figure gives a fair indication of the collective risk, doses to the gonads can produce genetic effects only if the individuals irradiated subsequently reproduce themselves. If the radiation is the same for everyone (as for example with cosmic rays—although there is some variation with latitude and altitude) then the genetically significant dose (GSD) is conventionally regarded as being equal to the average dose. But if some groups receive more than others, then a weighted average has to be taken, with the doses weighted by the individuals' expectancy of future children which declines with age.

45. The table gives only average doses, and conceals considerable variations. The highest doses will be received by the small numbers of people receiving intense radiotherapy, where doses of hundreds of rem to particular parts of the body are often given. Radiation workers are permitted to receive up to 5 rem/year as a whole-body dose, which is some fifty times the average exposure to naturally-occurring radiation. Amongst the general public there are variations because of medical diagnosis or treatment, and because of where they live. For example, granite contains a small amount of uranium and hence is very slightly radioactive. Accordingly, the level of terrestrial background radiation in granite-built houses in Aberdeen is about 150 mrem/year, whereas it is only 30 mrem/year in brick houses in London. In some parts of the world where deposits of radioactive minerals are found, the levels may reach 1–2 rem/year.

46. Of the man-made sources of radiation, some can be terminated at will such as medical and occupational exposures, but others are the ineluctable result of man's past activities. As a result of fallout from atmospheric bomb tests, we have all accumulated a very small quantity of radioactive strontium–90 with the calcium in our bones, and this will continue to irradiate our bodies for many years, although slowly decaying with a half-life of 29 years. We are thus committed to receive this radiation dose—estimated in 1972([4]) at 62 mrem—and a number of others from radioisotopes both within and outside the body. The nuclear power programme also gives rise to a number of very long-lived radioisotopes which will remain in the environment for many years, and in some instances, for millenia. It is therefore important to consider not only *current* annual doses from nuclear activities, but also the future "*dose commitment*" to radiation workers and to the general population that may be involved.

47. The radiation doses at present received as a result of nuclear power production are very small, and the great majority of the "collective dose", which is the summed total of individual doses and expressed in man-rem, is received by workers in the industry. The collective whole-body dose to the latter

was estimated(5) as 12,500 man-rem for 1974, of which two-thirds* was genetically significant. The same publication estimated the public dose as 500 man-rem, but a later publication(6) indicated that the effects of discharges to water were much greater than had been allowed for (see paragraph 80) and the total is more likely to be in the order of 1,900 man-rem all of which would be genetically significant. These figures imply that the total GSD to the general population as a result of the current level of nuclear power production is about 0·2 mrem/year, comparable with that from miscellaneous sources such as luminous watches, and about one-five hundredth part of what occurs naturally. However this average figure conceals very wide variations in doses between individuals. Some members of the public receive a few tens of mrem per year, and some radiation workers up to the legal limit of 5,000 mrem (5 rem).

Biological effects of radiation

48. Living tissue consists of cells, many of which have the ability to divide and so reproduce themselves. Very high doses of radiation can kill a cell, but radiation at much lower levels can prevent a cell from dividing, or damage the genetic material which is contained in its nucleus†, a separate structure in the centre of nearly all cells, and cause it to divide abnormally. The mechanism is thought to be the action of chemically active groupings of atoms which are formed as a result of the ionisation produced by the radiation, especially in the presence of oxygen. A cell is more likely to be damaged if it is irradiated while it is growing and dividing, and tissues containing many dividing cells are therefore called radiosensitive. Bone marrow, in which the blood cells are formed, is an example.

49. If a radiation dose is absorbed all at once, it may be more effective than if it is given as several smaller doses, or slowly over a long time. This is because there are certain natural repair mechanisms that permit cells to mitigate some of the effects of radiation. In practice, workers in the nuclear industry and members of the general public are likely to receive multiple small doses, from each of which there will be some recovery. However it is normal practice in radiological protection to ignore this effect, and to limit doses on the basis of estimated risks from single brief exposures.

50. *Long-term effects.* The principal long-term effect of radiation is the induction of cancers. Another effect that has been postulated is a non-specific increase in mortality from all causes. There is, however, no convincing evidence that such an effect has been shown in man by the doses with which we are concerned. Unlike some chemical carcinogens, radiation can induce cancers in any tissue of the body, but in some either earlier or more easily than in others. There is usually a delay of some years, or even decades, between irradiation and the appearance of a cancer. This makes it very difficult to be sure that any particular cancer is due to the radiation and it is necessary to use statistical methods to establish the connexion. Nevertheless, there is strong evidence that

* The factor arises because radiation workers have a greater average age than the general population and hence a reduced expectancy of future children.

† The cell nucleus is not to be confused with the atomic nucleus.

radiation has caused a number of additional cancers in the survivors of the Japanese atomic bomb explosions, the Marshall Islanders who were subjected to bomb-test fallout in 1954, uranium and other miners exposed to radon in the air at work, early radiation workers, and patients who have been treated with radiotherapy. There is also abundant evidence from animals, which is, in general, consistent with the observations that have been made on man.

51. There is some doubt about the minimum radiation dose needed to cause a cancer. Both direct animal experiments and epidemiological studies on man tend to be inconclusive because of the immense number of subjects required to determine with a high degree of confidence whether an increased incidence of cancer has occurred, and that it is not caused by some factor other than radiation. There is evidence of effects on the foetus at doses of a few rem, but it is possible that the foetus is more susceptible than the adult. Studies of the children of women who had received abdominal X-rays during preganancy have shown that exposure to 1–2 rem produces approximately a 50 per cent greater chance of dying of leukaemia or other cancer in the first ten years of life than occurs in the absence of such irradiation.

52. It is assumed in radiological practice and in the control of occupational radiation hazards that there is no threshold dose needed to induce cancer and that the effect of radiation is directly proportional to the dose down to the lowest doses and dose-rates. This assumed linear relationship is generally regarded as conservative, that is, as likely to lead to an over-estimate of the number of cancers that would be produced. It follows from this assumption that some risk is involved in any exposure to radiation, however small. Many studies have been conducted to assess the risk of fatal cancer induction from low levels of radiation; such work is clearly essential in order to assess the likely hazard to radiation workers and to the general public, so that the risk may be set against the expected benefits. A reasonable estimate of the number of fatal cancers that would be induced by a dose of 1 rem given to each of a million people would be of the order of 100, of which perhaps one quarter might be leukaemias. This means that radiation workers who receive an annual dose of 1 rem are running a risk of about 1 in 10,000 that they will eventually die of cancer as a result of each year's dose. This is approximately as dangerous as regularly smoking three cigarettes a week.

53. It has been estimated(3) that irradiation from medical X-ray examination (in which doses of about 100 mrem are given to bone marrow and of between 30 mrem and several rem to the skin) will give rise to one or two extra cases of leukaemia a year (compared with the spontaneous rate of 50–60 per year) for every million people involved. This risk is considered acceptable when balanced against the enormous medical benefits of diagnostic radiation. However, since it is apparent from Table 2 that medical radiation constitutes much the largest man-made source to which members of the public are exposed, on average, it is clearly the prime target for efforts to reduce overall exposure if this is considered to be excessive. We have not taken this subject as within our field of enquiry but we did receive evidence on it from the British Institute of Radiology and, believing the matter to be important, we have reproduced this in Appendix 1.

54. Just as natural radiation comprises γ-rays from outside the body and α- and β-particles from radioisotopes within the body, so the effects of radioactivity from the nuclear power programme are of both kinds. Nuclear fission gives rise to many different isotopes, and their biological effects depend upon how they are distributed between the different organs and how quickly they are eliminated from the body. Their behaviour or "metabolism" is therefore studied in ongoing research programmes in many countries. The difficulty is that it is naturally considered unethical to carry out experiments on man except with very small doses, and there are often doubts whether animal experiments provide a sufficiently close parallel because of variations between species in the way radioisotopes are dealt with by the body. Doubts have been expressed to us on this score about the results of some studies on rats which are extensively used for investigating the distribution of plutonium. Knowledge of species differences in metabolism is necessary to allow conclusions to be drawn on which organs are most at risk from particular radioisotopes, and consequently how much of the isotope may be allowed into the body without exceeding the permitted radiation levels; that is, the "maximum permissible body burden".

55. Some isotopes are eliminated rapidly from the body, and therefore their physical half-life alone does not convey a true idea of how quickly the radiation dose from a particular intake is reduced. For example tritium, a radioactive isotope of hydrogen with two neutrons in the nucleus, has a half-life of 12·3 years, but half of any quantity introduced into the body is removed within eight days. It is therefore not considered very radio-toxic except when it is incorporated in a few compounds that are retained in the body. On the other hand, strontium–90, as we mentioned above (paragraph 46), is chemically similar to calcium and may be incorporated in the bone structure where it remains for many years. Since its radiological half-life is fairly short (29 years) it is highly radio-toxic. The behaviour of a radioisotope may also depend upon its chemical form—whether it be soluble or insoluble—and upon how it is introduced into the body. Thus plutonium (a long-lived α-emitter, see paragraph 66) is extremely dangerous if it is inhaled but relatively harmless if swallowed as it is hardly absorbed through the gut wall and virtually all of it passes through the body.

56. The different parts of the body vary in sensitivity to radiation, and thus he allowable dose equivalent has to be related to the part exposed, either from a localised external source, or from the concentration of a radioisotope into a particular organ. The currently accepted limits are shown in Table 3. They derive from the recommendations of the International Commission on Radiological Protection (ICRP), which we describe in detail in Chapter V. The thyroid gland is unusual in that it tends to accumulate any iodine introduced into the body. This would include the radioactive fission products, iodine–129 and –131, were they to be encountered. The effects of an accidental release of large amounts of radioactive iodine on members of the public could be countered by their taking tablets containing stable iodate or iodine salts. This iodine would swamp the thyroid gland and prevent its uptake of more than a tiny fraction

of the radioactive iodine. There are in fact plans for the distribution of such tablets to the public in emergency situations, see paragraphs 218 and 255.

TABLE 3

ICRP recommendations of allowable radiation dose equivalents for radiation workers, per year. (Values are ten times those for individual members of the public* and thirty times those for exposed populations) (Published in 1966)

Organ irradiated	*Dose, rem*
Whole body Bone marrow Gonads	5
Skin and bone Thyroid gland	30
Hands and forearms Feet and ankles	75
Other single organs	15

* Except for the thyroid gland of children up to 16, where the limit is 1·5 rem/year.

57. *Genetic effects.* We have explained that the effect of radiation in cells is to interfere with their self-replication. In ordinary body cells, such interference can lead to the production of cancerous cells. But in the reproductive cells, sperm in men and ova in women, which are contained in the gonads, there is only a single set of chromosomes, and cell division and replication does not take place until after fertilisation. These chromosomes, which are made up of genes, carry the genetic information that determines the character of the offspring, and they can be damaged by radiation. Such damage may take the form of gene mutations or chromosome aberrations. Some of these changes result in offspring suffering abnormalities which may range from mildly detrimental to severely disabling or lethal disorders. The change in the genetic material may be dominant, in which case it will appear in the first generation of descendants, or it may be recessive or sex linked. A recessive mutation will only appear in a later generation, and then only if it is paired with a similarly mutated gene. Thus the absence of genetic abnormalities in the first generation is no guarantee that they will not occur subsequently.

58. We should mention that although the mutagenic properties of radiation are well known, there are also numerous chemical substances which are known to have a similar effect. The risk of genetic mutations from man-made radiation must be seen in relation to those occurring as a result of the natural sources of radiation listed in Table 2, and from other causes. Such genetic mutations are taking place all the time, and are a mechanism in the process of evolution since they provide a means whereby a species can adapt and survive in the face of changing environmental conditions. Ultimately the evolutionary process is beneficial for the survival of a species, but it takes place over many generations and for every beneficial mutation that occurs there are many that are damaging or deleterious. We are concerned with the health of individuals rather than the

long-term adaptability of our species and we must regard additional mutations as harmful, but they have to be seen in perspective with other environmental hazards.

59. Radiation-induced genetic effects were observed in insects in 1927 and have been produced in animals under experimental conditions, but no quantifiable effects have been established as having occurred in children born to survivors of the Hiroshima and Nagasaki bombs.* There is no reason to suppose, however, that man is immune from genetic damage from radiation and the present practice is to estimate the effect of a given dose on man by extrapolation from results obtained with flies and mice. Here again a linear hypothesis is normally used, although it is known that doses given over a short period are more damaging than if they are spread over a long time, implying some genetic repair mechanisms.

60. It should also be pointed out that a number of mutations are so serious that the foetus is not viable, and will abort spontaneously, very probably at an extremely early stage of pregnancy when the event may not be recognised as such. This already happens very frequently as a result of naturally-occurring genetic flaws. Still others may produce variations of such minor concern that they would have no discernible effect on the individual.

61. The amount of genetic damage from a given dose of radiation can be assessed either by an analysis of the number and type of mutations that occur in the genes and chromosomes, or from estimates of the dose needed to double the number of naturally-occurring genetic defects. Such estimates vary in the range 10–100 rem. If a population of one million people were each to receive a genetically-significant dose of 1 rem over a generation of 30 years (the legal limit in the UK), then on the basis of figures given in both the UNSCEAR(4) and BEIR(9) reports, there would be about 10 substantial genetic abnormalities per year in the ultimate descendants, with between one and two per year in the first generation. The incidence of genetic effects per unit of radiation dose is less well known than is the incidence of cancers, and there is a factor of uncertainty of about five on these estimates, which could be either way.

62. The naturally-occurring incidence of genetic disease is such that it affects some 6 per cent of all babies, so that a population of a million people with an annual birth rate of 12 per 1,000 would in any event generate some 720 cases each year. The additional incidence from radiation doses equal to the legal limit would most likely be far less than the normal statistical variations in this number and would therefore be imperceptible. Moreover such an average radiation dose would be a significant fraction of the natural level, and we would expect that it would have been prevented from reaching such a

* Although it was previously considered that people (and rats) living in areas of high natural background radiation suffered from no significant mutation effects, a recent paper(8) has shown that there has been an increase in severe mental retardation and in chromosome aberrations in children born to mothers in Kerala, South India, where background radiation doses are between 1·5 and 3 rem/year. The incidence is particularly high among children born to mothers over 30, who would have received higher doses.

limit long before in order to keep the numbers of somatic effects at a low figure. We are therefore satisfied that in these circumstances, genetic effects should be of little concern.

63. *Acute effects.* We have been considering the long-term effects of the small doses of radiation to which members of the public and radiation workers may be exposed as a result of the routine operation of the nuclear power industry. In extraordinary circumstances, such as if an accidental release of radioactivity occurred (or, in the most extreme case, if a nuclear weapon was exploded) much larger radiation doses could be received. We therefore end this section with a brief description of the acute effects produced by large doses.

64. One way of describing the effects on different species is to tabulate the radiation dose that would kill half the population exposed within a specified time, as in Table 4.

TABLE 4

Radiation doses required to give various species a lethal dose to 50% (LD 50) within 30 days

Pig	200 rem	Gerbil	1,000 rem
Man	250 rem	Snail	10,000 rem
Mouse	650 rem	Fruit fly	60,000 rem

These are only approximate values, and for both man and other species the lethal dose varies considerably, depending on age and physical condition. For obvious reasons there are particular uncertainties about the figure shown for man but it appears that man is more radiosensitive than most species so far tested. In general the more complex the animal, the lower the dose of radiation required to kill, though as the Table shows there may be exceptions.

65. The probability of death increases with the dose received, and a dose of 500 rem in man would probably be fatal. Doses of about 500–1,000 rem would normally kill a man within a few weeks because of the failure of the bonemarrow cells to make the fresh blood cells that are required to enable the body to resist infection. Doses of about 1,000–5,000 rem cause failure of the gastro-intestinal tract, extensive internal bleeding, and death within a few days. Even higher doses, say greater than 10,000 rem, lead to a failure of the central nervous system, with seizures and convulsions, and death within a day. If high doses of radiation are given only to parts of the body, other forms of damage may occur. Radiation on the skin produces effects rather similar to those of burns. The hair-forming cells are radio-sensitive, and surface doses of 300–400 rem lead to loss of hair, which becomes permanent at twice this dose. Sterility results from irradiation of the gonads, and even doses as low as 10 rem can lead to a temporary loss of fertility in men.

The biological effects of plutonium

66. We now consider the biological effects of one particular radioactive substance, plutonium, in more detail. This will illustrate the kind of considerations involved in assessing the effects of any radioisotope in man. We have

chosen to discuss plutonium because it is exceedingly toxic and because it is produced in substantial quantities in a nuclear power programme. There are 16 known isotopes of plutonium but we are principally concerned with plutonium–239, which has a very long half-life (24,400 years) and, like all plutonium isotopes, is retained in the body once it gains admittance (except by ingestion as mentioned in paragraph 55). Its behaviour in the body depends on how it enters and whether it was originally in a chemically soluble form, such as the nitrate salt, or an insoluble form, such as the dioxide. Some of the other isotopes of plutonium, such as plutonium–238 (half-life 86 years), being shorter-lived have much higher specific activities, and are observed in animal experiments to translocate much more readily within the body than would oxide particles of plutonium–239, so they may be considered as moderately soluble.

67. The ICRP recommend basic radiation standards (see Table 3) for radiation workers and members of the public, in the form of maximum permissible annual doses to the whole body and to particular organs. For practical purposes it is necessary to convert these into secondary standards for plutonium (and other radioisotopes) which specify the maximum permissible annual intake and the maximum permissible body (or organ) burden. This requires a knowledge of how a given intake of the element will be distributed among the different body organs, how the distribution will vary with time, and how effective the deposited plutonium will be at causing damage in the tissues or in producing a cancer. Plutonium forms chemical organic complexes, and finds its way into a number of organs, not all of which are yet generally recognised as specific target organs. It is thus important in any animal studies or in any survey of people exposed to the element that account should be taken of any abnormal incidence of cancer or other disease even if it is of a tissue not previously thought to be specially at risk.

68. Small, insoluble particles that are inhaled tend to remain in the lung, although there is normally some movement towards the lung periphery and the thoracic lymph nodes which tend to receive a higher dose. It has been ICRP practice to average the radiation dose received over the whole mass of an organ unless there is clear evidence that this assumption is not conservative. Two years ago, and indeed just as our study was beginning, it was suggested([10]) that small "hot" particles of plutonium dioxide giving a radiation dose to the immediately surrounding tissue of 1,000 rem/year might be very much more carcinogenic than a similar amount of plutonium uniformly distributed over the lung volume. Since then, further articles have appeared([11]) in support of this suggestion and it has been the subject of some public discussion. In the UK, the National Radiological Protection Board (NRPB)([12]) and the Medical Research Council (MRC)([13]) have both published reports in which they conclude that the suggestion is invalid.

69. We appreciated that if it were true that "hot" particles represent an additional hazard, this would have profound implications for our study, because it might require a reduction in exposure to plutonium to levels that would be difficult to detect and control, and effectively preclude the use of nuclear power.

We therefore engaged independent consultants to examine this issue and to advise us on plutonium toxicity generally. Their findings support the contention of the MRC, namely, that "there is at present no evidence to suggest that irradiation of the lung by particles of plutonium is likely to be markedly more carcinogenic than when the same activity is uniformly distributed". They also cited measurements of bones of patients whose radium burdens had caused bone cancers as additional evidence that, where the number of cells affected by radiation was relatively low, the carcinogenic risk was also small. As a result of our review, we are satisfied that no case has been made in favour of the special carcinogenicity of "hot" particles. Indeed, there is strong circumstantial evidence to the contrary. The suggestion does not, therefore, provide a reason for dramatically tightening standards beyond those that are now generally accepted, although adjustments by small factors will no doubt continue to be made from time to time. Nevertheless the problem is so important that we would like to see further experiments mounted to confirm the correctness of the current conclusions.

70. Another suggestion that the risk of cancer from inhaled plutonium particles might be much greater than previously supposed has been advanced by Gofman([14]), based on consideration of the parts of the lung at risk and on the effects of smoking on lung clearance rates. We arranged for this suggestion also to be reviewed by the NRPB, the MRC and our consultants. They all independently agreed that the arguments in the papers were erroneous and that no further revision of standards was necessary as a consequence. Our call for detailed review of material that had not been published in a journal of scientific standing (when it would have been subjected to examination by competent referees) provoked a comment that such work was time-consuming and could be counter-productive in diverting attention from the real issues of plutonium dosimetry. We note here, however, that because of the emotive nature of the issues raised by plutonium, which concern many people who are unable themselves to assess the scientific arguments, much attention is focused on work that suggests that current assessments of the hazards are seriously wrong. It appears to us desirable that work arousing serious public concern should indeed be examined and, if appropriate, publicly refuted on the basis of more serious arguments than the mere denigration of the authors.

71. We have seen that so far as effects on the lung are concerned, plutonium-239 in insoluble form is of most significance. The current "maximum permissible lung burden" (MPLB) of 0·016 μCi, or 0·25 microgram, is based on averaging the dose from inhaled plutonium over the whole lung mass, to give a dose equivalent of 15 rem per year for a radiation worker. Studies in the USA of a small number of plutonium workers with lung burdens approaching or exceeding this level have not disclosed lung cancers. Accident figures from the USA indicate that inhalation is the main route of entry and the possibility of a fire involving plutonium which could disperse it in particulate form is a matter of concern. The follow up of accident cases, as is proposed in the USA Trans-uranium Registry, will enable better estimates of the margins of safety operating under either accidental or routine exposure conditions to be established.

72. The potential radiotoxicity of plutonium was recognised when it was first produced by Glenn Seaborg over 30 years ago. It was found that the main sites for the deposition of plutonium entering the body in the form of soluble compounds were the bones, including the bone cells lining the bone marrow cavity. Because of this bone-seeking property it was possible to derive a "maximum permissible body burden" (MPBB) by comparison with the body burden already derived for radium–226. This was founded on well-established data relating the ingestion of this radioisotope and the amount retained in the body at death to the incidence of bone cancers decades later. Radium–226 disperses through the bone volume whereas plutonium mainly collects on the bone surface. Therefore in deriving the MPBB for plutonium a factor of five was introduced to allow for this difference in distribution; a further factor of two or five could be introduced to allow for the higher toxicity of plutonium–239 compared with radium–226, as established from experiments with dogs. The current MPBB for plutonium–239 is 0·04 μCi, or 0·6 microgram. Our consultants confirm that this is a reasonable level for allowable skeletal deposition, and that the much lower level recently proposed by Morgan([15]) cannot be substantiated([16]).

73. It has been realised since 1947 that plutonium accumulates in the marrow where the blood-forming cells are to be found as well as on bone surfaces. The ICRP have in fact suggested([17]) that the skeleton should be treated as if it consisted of two separate organs when the metabolism of plutonium is being considered. The possibility of a consequent leukaemic risk was pointed out in 1967([18]) but little attention has been paid to it. Indeed the MRC report on plutonium toxicity([13]) barely alluded to it. Early in 1975, public attention was drawn([19]) to the fact that leukaemia had developed in several of the workers at Windscale and other plants where plutonium is extracted or fabricated. BNFL and the NRPB denied that any deaths attributable to plutonium had occurred, but we felt that the facts needed to be clearly established and we asked the NRPB to do this and set them out for public examination.

74. In September 1975 the NRPB produced a paper([20]) in which a comparison was made between the numbers of observed and expected deaths at Windscale alone from leukaemias and other types of cancer including those involving the bone marrow. They concluded that the excess of observed deaths was not statistically significant and appeared to regard the matter as closed. Yet the data they reported were limited to observations on men whilst they were actually employed and were therefore estimated to have omitted at least 50 per cent of the deaths from cancer that would have been found had the men continued to be kept under observation after they had left employment. It is a common experience in industrial medicine to find that observations limited to the period of employment are biased by a deficiency of deaths from cancer and chronic diseases and it is difficult to understand why it has not been possible to carry out a proper study of all radiation workers, whether or not they have ceased employment, as has been commonplace in other industries.*

* It may be noted that any leukaemias from plutonium would be expected to have developed after a longer time than usual because of the slow rate of translocation of plutonium to the bone marrow.

75. On the basis of the information that we have been given about the doses of radiation that the men are likely to have received, we are not persuaded that any notable hazard has existed. Nevertheless, it is important to be sure that it has not, particularly in view of the findings of an excess of leukaemia in patients who had been given thorotrast† which was appreciably greater than the expected excess of bone tumours([21]). We have, therefore, asked the Board to seek out all the records (which we understand are available) to allow an analysis to be made of the mortality from all cancers, including, in particular, those of bone marrow origin, among all the employees of the nuclear industry, both during and after employment, who might have been exposed to plutonium.

76. A substantial amount of research work on plutonium toxicity is already being carried out, but this is not excessive in view of the extent to which plutonium may be used in the future. Some of the work, such as that done at the NRPB, is directed towards the treatment of people who accidently become contaminated. We are glad to see that these studies are being extended to other animal species and not just rats. We welcome the effort that the NRPB is now putting into establishing a National Registry of Radiation Workers covering all those employed at the beginning of this year, which will eventually show whether any anomalously high death rates occur which need to be further investigated. We consider, however, that it is essential that all ex-employees should be included so that the Registry will be complete. Radiation records must be statutorily retained for 30 years, so this would be feasible. We discuss this again in paragraph 253.

77. Subject to anything that may emerge from further analysis of cancer incidence at Windscale and elsewhere, we do not consider that the derived standards for plutonium exposure and uptake are seriously in error. The MRC have pointed out that the maximum permissible concentration for insoluble particles in air is too high by a factor of 5, but this has been recognised and we were informed by the NRPB that the nuclear industry had been applying the more stringent standard for some years.

Movement of radioisotopes in the environment

78. As a result of the medical studies on the effects of radioisotopes on man and on animals, there has been drawn up a list of the "maximum permissible concentrations" (MPCs) of individual isotopes in air and water which correspond to the ICRP recommended maximum dose levels. These are used for checking the environment of radiation workers, and are based on exposure during a 40-hour week, so that a worker subjected to the particular concentration would be receiving the maximum permitted dose to a particular organ or to his whole body. For example, in drinking-water the concentrations range from 0·2 curies/m^3 for tritium (on a whole body basis) to 4 microcuries/m^3 for strontium–90 (on the basis of its concentration in bone). However radioisotopes do not usually remain static in the environment, and they may be concentrated

† Thorotrast is a thorium-based contrast medium formerly used with X-rays.

in unexpected ways through the biological food chain or otherwise. To decide whether it is safe to dispose of radioactive waste, it is not enough just to check that it will not exceed the permitted maximum concentrations in the air or water to which it is discharged; it is also essential to check all the possible "environmental pathways" to ensure that particular individuals or groups, such as fishermen whose nets may be contaminated, are not subjected to undue risk.

79. For any proposed release of radioactivity to the environment, it is usually found that there is one particular pathway whereby a group of people (the "critical group") are given a radiation dose which is much larger than that received by the rest of the population. If members of this group are considered to be receiving an insignificant dose, then it follows that all other members of the public will also. The group may on occasion be very large, in which case estimates are made of the integrated dose to the total membership and account is taken of the need to limit the dose for genetic reasons. It may consist of just a few people or even on occasion a single person whose particular dietary and working habits are studied in meticulous detail.

80. An important feature of critical group analysis is to recognise that the environmental pathway of greatest radiological significance may change with time for a given discharge of radioactivity to the environment. Thus, for the discharge to the Irish Sea from Windscale on the Cumbrian coast([6]), the limit used to be determined by the concentration of ruthenium–106 in an edible seaweed, *Porphyra*, which was gathered along the coast and sent down to South Wales to be made into laverbread, a delicacy normally fried and eaten with bacon. The MAFF carried out dietary surveys and identified a few individuals with a particular liking for laverbread; these formed the "critical group". Recently the *Porphyra* gatherers in Cumbria have retired and laverbread is made with seaweed from elsewhere which contains hardly any radioactivity. The critical group then became a few salmon fishermen in the Ravenglass estuary, who received external exposure from γ-radioactivity deposited on the shore. The increasing discharges of caesium–137 have now led to higher concentrations in fish and to a significant population dose to the large number of people who eat fish and shellfish from the Irish Sea. It is unlikely that this will be the last change in the critical group, and there is clearly a continuing need for an extensive programme of work on the movement of radioisotopes through the environment, and how they are distributed through living organisms. We describe the arrangements for this work in more detail in Chapter V.

Effects on the natural environment

81. There is a general presumption that if radioactivity discharges are limited by consideration of human health, then the natural world will be unaffected. It is known that some animals are more sensitive to radiation at certain stages of the breeding cycle and that they may selectively concentrate radioisotopes in a similar manner to heavy metals and persistent organic chemicals. It may be therefore that significant numbers of animals and plants suffer radiation

damage, and show a loss of fertility and an increased incidence of tumours. In the natural world, however, we are mainly concerned not with the health of individuals but with whether a species is endangered; and public attention is concentrated on attractive and conspicuous animals and plants. Populations in the wild are usually limited by density-dependent factors such as food supply or the action of predators and parasites. Minor changes in fertility and survival caused by radiation are therefore unlikely to disturb the equilibrium of populations. But it is possible that radiation could tip the balance against a species that was otherwise hard-pressed to survive in a particular habitat. In some cases also it is important to preserve the existing population of a species as well as the species itself. Even if the English elm survives as a species, the current reduction of the population by Dutch elm disease is an environmental disaster. There is no reason to expect an equivalent disaster from radiation doses within the current limits, but ecological changes are sometimes unforeseen and there is a case for maintaining a watch on the ecology of areas which receive higher radiation doses.

CHAPTER III

NUCLEAR POWER

Introduction

82. In this chapter we describe the main features of nuclear reactors and of the nuclear fuel cycle. We begin with a simple presentation of the physical principles on which reactor operation is based, and give a brief account of the different types of reactor that are in use or planned. The term "nuclear fuel cycle" refers to the totality of operations involved in the fabrication and treatment of nuclear fuel. It includes the movement of uranium from mine to fuel fabrication plant and thence to reactors, the removal of spent fuel and its treatment at the fuel reprocessing plant to extract material suitable for incorporation in fresh fuel, and the treatment and disposal of wastes. We describe this cycle in some detail, for it is important to an understanding of the radiological hazards that can arise.

Nuclear physics

83. In paragraph 27 we explained that atomic nuclei are made up from protons and neutrons ("nucleons"), and that binding forces overcome the mutual repulsion between the positively charged protons. If a nucleus could physically be separated into constituent nucleons, it would be found that the mass of the nucleus would be a little less than the sum of the masses of the individual free nucleons. In coming together to form the nucleus, the particles have given up a small amount of mass or, since Einstein's famous relation $E = Mc^2$ shows that there is an equivalence between mass and energy (the conversion factor being the square of the speed of light), a certain amount of energy.

84. This energy is called the "binding energy", and atoms in the middle of the periodic table are more strongly bound than are atoms at the beginning and end. Figure 3 shows the curve of binding energy. The shape of this curve suggests two possibilities for releasing energy. If two very small nuclei are joined together or "fused" then the new composite nucleus will be more strongly bound than its components, and the difference in energy will be given out in the process of fusion. It is this energy that powers the sun, in which hydrogen, the lightest of all the elements, is fused into helium and heavier elements. However fusion can take place only under extreme conditions of temperature and pressure such as are found in the interior of the sun and other stars. They have been reproduced on earth in the hydrogen bomb. Attempts are being made to harness this energy in a controlled way in a fusion reactor and we revert to this subject at the end of this chapter (paragraphs 145–152).

85. The other possibility suggested by the curve of binding energy is that if a very heavy nucleus could be made to divide or "fission" into two parts, each of these would be more strongly bound than the original nucleus and

FIGURE 3

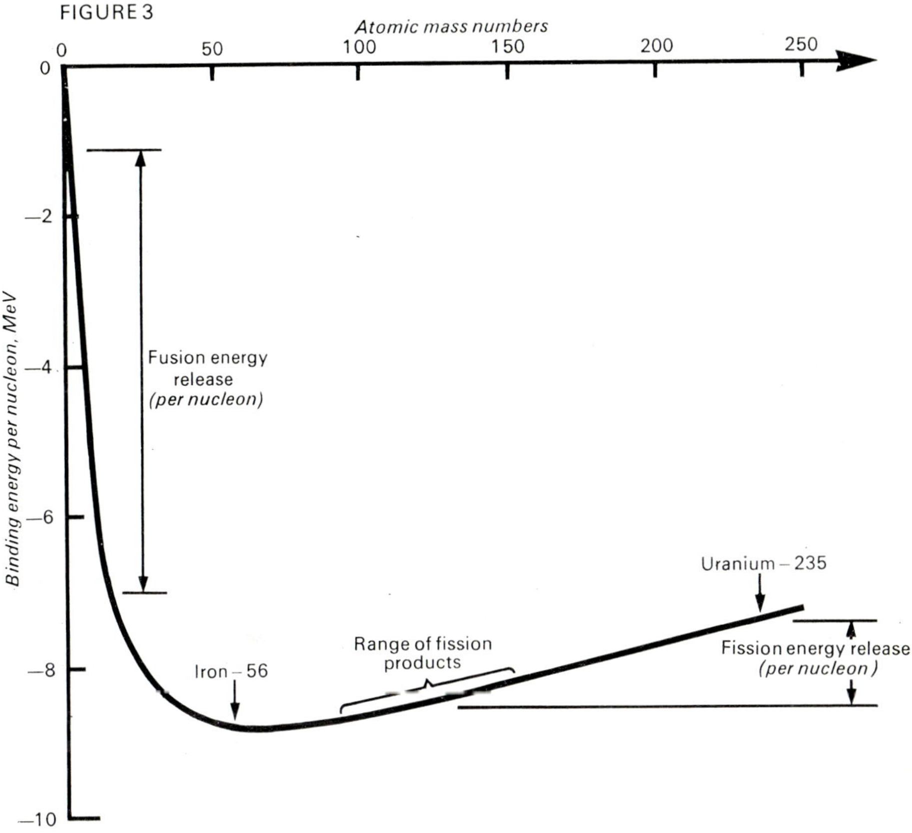

The curve of binding energy.
(1 MeV/nucleon ≡ 2,68 × 10^7 kWh per kg.)

again the energy difference would be released. This was shown to be theoretically possible in the 1930s, and in 1945 it was dramatically demonstrated by the first atomic bombs. The remarkable amount of energy made available through fission may be gauged from the fact that the energy released by the fissioning of one kilogram of uranium–235 would be equal to that obtained by burning about 3,000 tonnes of coal. In nuclear reactors the fission process has been controlled so that the energy release takes place steadily and continuously.

Fissile materials

86. A few radio-isotopes are so unstable that they fission spontaneously, although none that occurs in nature does so to a significant extent. However, some radio-isotopes have the property that their nuclei may be induced to fission if they first capture neutrons. Such substances are called "fissile" and one of these that occurs in nature is uranium–235 which is present in natural uranium to the extent of about seven parts per thousand. If a nucleus of

uranium–235 is struck by and captures a neutron, and then fissions (and if it has captured a neutron, the chance of it fissioning is about 86 per cent), it will form two "fission products", together with two or three surplus neutrons. (See paragraph 31.) The fission products will be radioactive isotopes of various lighter elements, depending on how the uranium nucleus divides. They might, for example, be tin and molybdenum. The energy released by the fission appears mainly as kinetic energy of the fission products and the spare neutrons. If one of these neutrons should strike and be captured by another fissile nucleus, the process can become a self-sustaining chain reaction.

87. Some nuclei which capture neutrons will not themselves fission readily but they will change by radioactive decay into another nucleus that is fissile. Thus uranium–238 captures a neutron to become uranium–239, which quickly decays by two successive β-emissions to become neptunium–239 and then plutonium–239. None of these three occurs in nature in significant amounts because their half-lives are so much shorter than the age of the earth and they do not have a long-lived radioactive parent. The last isotope is of extreme importance because it is fissile, and because it can be "bred" by this process from the much more abundant uranium isotope, uranium–238. There are a number of other fissile materials, but uranium–235 and plutonium–239 are the ones with which we shall mainly be concerned.

The chain reaction

88. For the fission process to be self-sustaining exactly one of the 2–3 spare neutrons from each fission must on average initiate another fission. If an assembly of pure fissile material is so arranged that on average slightly less than one neutron from each fission causes another nucleus to fission, then the fission process will stop. Such an assembly would be described as "sub-critical". But if the assembly is such that the average number of neutrons producing further fission is slightly greater than one, then the process will grow at an accelerating pace. In fact, it would accelerate very quickly indeed, as the average time between successive fissions is a minute fraction of a second. This assembly would be "super-critical", and in suitable circumstances it could result in an explosive release of energy. This is the principle behind the atomic bomb.

89. In a nuclear reactor, however, the designer controls the "reactivity" of the nuclear material so that the average number of neutrons, and therefore of fissions, in a given time interval remains constant. The assembly will be just critical, and the fission process will release energy in an orderly manner. Since the time required for a neutron ejected from a fissioning nucleus to cause another fission is exceedingly small, it might be thought that it would be impossible in practice to control the process so that it was consistently maintained at the critical level. Fortunately, however, not quite all the neutrons produced are released immediately. A few per cent emerge belatedly from the fission products with delays of the order of tenths of a second. A reactor is arranged to be sub-critical on the "prompt" neutrons, but to be critical when account is taken of the extra delayed neutrons. Under some circumstances it

could be critical on the prompt neutrons alone, or "prompt critical", but normally the approach to this condition would be detected by instruments and the reactor shut down long before it was reached.

Critical assembly

90. Criticality depends upon a number of factors, notably the concentration of fissile nuclei in the assembly, its size and shape (if its surface-to-volume ratio is too great, too many neutrons will escape from the surface), whether it is surrounded by a neutron-reflector, and whether the neutrons are travelling at the right speed readily to be captured by fissile nuclei and so initiate further fissions. In its simplest form a critical assembly may be a piece of fissile material. The optimum shape is a sphere which gives the lowest surface-to-volume ratio and so minimises the escape of neutrons from the surface. For any particular fissile material there is a certain minimum radius for a sphere that will just allow a steady chain reaction. The quantity of material required is called the "bare sphere critical mass", and for uranium–235 it is about 50 kg. Natural uranium which, as we have noted, contains the fissile isotope uranium–235 only to the extent of seven atoms in every 1,000, cannot by itself form a critical assembly no matter how large the piece, since the concentration of fissile nuclei is too low to overcome the capture of neutrons by uranium–238.

91. Neutrons escaping from the nuclear assembly may be reflected back to it if the surrounding material is suitable. This fact is used in the design of nuclear assemblies in order to prevent the waste of neutrons and thereby to reduce the amount of fissile material needed below the bare sphere critical mass. The minimum quantity of, say, plutonium needed to make an atomic bomb depends on the design adopted and the materials available, but it is considerably less than the bare sphere critical mass and may be as low as 4 kg which could be contained in a sphere 80mm in diameter([22]).

92. Fissile nuclei are much more likely to capture a neutron and subsequently fission if the neutron has a relatively slow velocity, comparable with that of molecules in a gas at moderate temperatures. Since this velocity reflects the thermal state of the gas, these relatively slow neutrons (actually they are travelling at a few km per sec) are called "thermal". They can be slowed down from their initial speed by means of a series of collisions with light nuclei. Not all light nuclei are suitable as some, such as boron, tend to capture the neutrons rather than slowing them down. The best nucleus, because it is both very light and has the least chance of capturing them, is that of deuterium, or hydrogen–2; others that are currently used are ordinary hydrogen, oxygen and carbon. These materials are called "moderators": they moderate neutron velocity, and by so doing enhance criticality. The use of a moderator enables a critical assembly to be achieved with a lower concentration of fissile material. In particular, using a moderator a critical assembly can be obtained even with natural uranium and some reactors are of this type. Other types of reactors are designed to use uranium which is enriched to varying degrees, that is, the proportion of uranium–235 is increased. We describe the enrichment process

in paragraphs 123–5. Absorbing substances like boron are used to control the numbers of neutrons in the assembly, and hence its "reactivity".

93. We have described the factors that are relevant to the construction of a critical assembly. It should perhaps be noted, however, that such an assembly may be created by accident, for example by the accidental juxtaposition of appropriate masses of fissionable materials during processing or storage. The action of moderators must also be borne in mind in this context. Thus, a store of fissile material, normally sub-critical, might become critical if it was flooded with water which would act as a moderator. In such "criticality incidents" there would be much γ and neutron radiation which could be very dangerous for personnel, and elaborate precautions are taken in nuclear plants to prevent their occurrence.

94. We are not concerned with the design of nuclear weapons but it is convenient at this point to mention some general aspects of this subject which are relevant to our later consideration of security issues. In the design of nuclear weapons, fissile material is used to create an assembly that is unmoderated and in which the fast neutrons alone create fissions. Thus although slightly enriched uranium (say, 4 per cent uranium–235) can form a critical assembly if a moderator is present it cannot be used to make a bomb because in such circumstances the chance of a neutron causing fission in uranium–235 rather than being absorbed in uranium–238 is insufficient to sustain a chain reaction. To make a bomb, the uranium must be highly enriched, normally to above 90 per cent. Such uranium, which can be made into a fast critical assembly, is referred to as "weapons grade".

95. The other main fissile material, plutonium, also exists in the form of several isotopes, and there is also "weapons grade" plutonium. This depends upon there being only a small concentration of the isotope plutonium–240.* Now this isotope is spontaneously fissile (see paragraph 31) and emits neutrons in the process. In the construction of a plutonium bomb, a sub-critical mass of material is compressed into a super-critical mass by implosion, and for maximum yield it is necessary for the chain reaction to be initiated at the instant when the core reaches its most reactive state. If there were a significant proportion of plutonium–240 present, the chain reaction might be started prematurely by a neutron arising from spontaneous fission and the energy released would then be considerably reduced. Weapons of comparatively low (and uncertain) yield can nevertheless be made from "civil grade" plutonium containing appreciable concentrations of plutonium–240. Such material is produced in civil nuclear reactors such as we shall be discussing.

Nuclear reactors

96. Anyone who visits a nuclear power station can hardly fail to be impressed by the nature and scale of the enterprise and by the technical achievement that

* Plutonium–239 is produced in a reactor (see paragraph 87), and although it is fissile on catuture if a neutron, there is also a significant probability that plutonium–240 will be formed instead. In its turn, plutonium–240 captures neutrons to form plutonium–241 which, like plutonium–239, is fissile.

it represents. Though the origin of nuclear power appears mysterious, however, it is important to appreciate the many parallels between the use of nuclear fuel and conventional fossil fuel for electricity generation. Thus, both processes involve the mining of basic materials (coal, oil or uranium ore), their processing and transport to the power station for combustion and the production of heat and its extraction to raise steam to drive turbo-alternators which generate electricity. Although nuclear fission is not strictly a combustion process, it is common to speak of the uranium being burned—it combines with neutrons rather than with oxygen—and both processes lead to the production of wastes, including waste heat, which have to be disposed of or put to other use. This is not, of course, to minimise the major differences that do exist, and especially those resulting from the hazardous nature of nuclear materials and wastes.

97. The main features of a nuclear reactor can readily be understood in the light of the principles we have described. A reactor is an assembly of nuclear fuel elements containing fissile material with means for the control of criticality so that the rate of fission can be regulated. Heat is generated through the ionisation caused in the fuel as the fission fragments are slowed down, and this heat is carried away by a coolant and used to raise steam. Figure 4 shows the main component parts. The fuel "pins" (see Plate 4) are clad in an impervious metal to retain the fission products, which accumulate within the uranium and are intensely radioactive. Surrounding the fuel elements is a moderator;

FIGURE 4

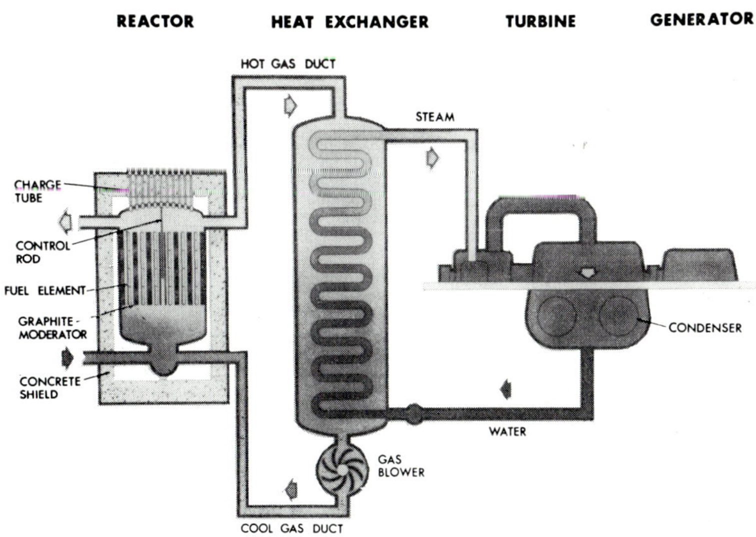

Schematic diagram of a thermal reactor (Magnox).

United Kingdom Atomic Energy Authority.

this may be a solid (carbon) or a liquid, heavy water (deuterium oxide) or ordinary water. The heat generated in the fuel is removed by a coolant, either a gas (helium, carbon dioxide or air) or a liquid (heavy or ordinary water, sometimes the moderator fulfilling a second role). In most reactor designs a heat exchanger is used to extract heat from the coolant and transfer it to water to raise steam for the turbines. We have explained that the operation of a reactor is based on neutrons. The means of control is usually but not always a set of rods made of boron steel which absorb neutrons and can be inserted quickly to shut the reactor down if needed. Since the core of a reactor is intensely radioactive, it has to be surrounded by heavy "biological shielding" to protect the operators from the neutrons and γ-radiation. This is usually concrete, several metres thick. Reactors of this kind, where there is a moderator and the majority of fissions are caused by thermal neutrons, are called "thermal reactors". Virtually all reactors now operating on a commercial scale are of this type.

98. The fission products produced consist of an enormous range of isotopes, including most of the elements in the middle of the periodic table (atomic numbers 33–64). Because of the high proportion of neutrons in the uranium nucleus the fission products tend also to be neutron-rich, and therefore to be β-emitters. Some of them are solids, but others are gases such as krypton, xenon and iodine (which is gaseous at reactor operating temperatures). These gases can exert considerable pressure on the fuel cladding, which must therefore be made to extremely high standards. As the fuel elements are irradiated, uranium–235 atoms are destroyed by fission and some of the fission products tend to absorb neutrons. Thus, the reactivity of a fuel element decreases with time to the point where it is said to be "burned-up". When a reactor is first started the new fuel elements contain surplus reactivity which is countered by the presence of the control rods or burnable poisons* within the fuel. As irradiation proceeds (this is measured by the amount of heat extracted in megawatt-days per tonne of uranium) the control rods are slowly withdrawn. Eventually it is necessary to replace the fuel elements with fresh ones in order to sustain the chain reaction and to prevent leakage of the accumulated fission products through the cladding. This is normally done piecemeal and with some designs can be done while the reactor is in operation.

99. We have seen that the contents of a reactor are extremely radioactive. An accidental release of radioactivity could constitute a serious hazard to people downwind of the reactor and would contaminate the ground where radioactive material was deposited. The severity of dose effects would of course depend on the magnitude of the release, on the weather conditions prevailing at the time and on the population density in the area affected. Safeguards are incorporated in reactor design with the aim of ensuring that the risk of accidental release of radioactivity is exceedingly small. Nevertheless, it is possible to postulate failure conditions that could lead to such releases. We consider the question of reactor safety further in Chapter VI but it is convenient to mention briefly at this point the major potential failure mechanisms.

---

* Isotopes that absorb one neutron readily but then decay to isotopes that do not absorb neutrons.

100. Clearly the control of reactivity is basic to safe operation of a reactor. A control failure leading to a rapid rise in reactivity, and hence in the fuel temperature, could result in melting of the fuel, which would slump to the bottom of the reactor vessel. It is important to emphasise that in a thermal reactor the fuel is near to its most reactive state in its normal arrangement, and there are very few possibilities that a core melt could lead to a critical assembly. However, the molten fuel could melt through the containment, and there is also the possibility (depending on the reactor type involved) that the containment could be breached by explosive forces generated by thermal or chemical interaction between the fuel and the coolant. This could result in the release of some of the volatile fission products, notably iodine.

101. In some types of reactor, melting of the core could occur as a result of a failure to remove the residual decay heat from the fuel after the reactor has been shut down. This heat is generated by the decay of radioactive fission products within the fuel; typically it amounts to about 6 per cent of the heat being removed when the reactor is at full power, but falling to about 1 per cent after an hour since much of the activity is associated with very short-lived isotopes. Thus, if a failure were to lead to a loss of coolant it would not be enough simply to shut down the reactor, and an independent, emergency core cooling system has to be provided to remove the decay heat. There has been controversy over the reliability of such systems in Light Water Reactors.

Types of thermal reactor

102. There are a number of different designs of thermal reactors, using different types of fuel, fuel cladding, moderator and coolant, and they are referred to by acronyms of bewildering variety. Although not specifically concerned with the differences between them we thought it would be helpful to give a very brief description of each type with its salient features. The main characteristics are summarised in Table 5.

TABLE 5

Characteristics of major thermal reactor types

Reactor	Magnox	AGR	CANDU	LWR	SGHWR	HTR	Water/ Graphite
Country of Origin	Britain France	Britain	Canada	USA	Britain	Several	USSR
Fuel	Metal Natural	Oxide 2 per cent enriched	Oxide Natural	Oxide 3–4 per cent enriched	Oxide 2–3 per cent enriched	Carbide 93 per cent enriched	Oxide 1·8 per cent enriched
Cladding	Magnox	Stainless Steel	Zircaloy	Zircaloy	Zirconium	Silicon Carbide	Zirconium/ Niobium alloy
Moderator	Graphite	Graphite	Heavy water	H_2O	Heavy Water	Graphite	Graphite
Coolant	CO_2	CO_2	Heavy Water	H_2O	H_2O	Helium	H_2O
Fuel burn-up, MW-day/ tonne	4,000	18,000	9,000	20,000	21,000	100,000	18,500
Power-density, kw/litre	1	4·5	16	50–100	11	6	n.a.

103. Magnox reactors were developed in the UK in the 1950s, and a substantial number have been built; in fact until very recently virtually all our nuclear electricity (amounting to an installed capacity of 4,200 MW) was generated in power stations using this type of nuclear steam supply system. The fuel is natural uranium metal, cast and machined into cylinders about 1 metre long and 25 mm diameter, and encased in a magnesium alloy ("Magnox"). The moderator is of carbon and the core consists of a large number of graphite blocks in which holes have been machined to receive the fuel elements, the control rods, and instrumentation. The coolant is carbon dioxide under pressure. The early designs have heat exchangers external to the reactor pressure vessel which is a steel sphere, with walls between 75 and 104 mm thick. The last two, at Oldbury and Wylfa, have pre-stressed concrete cylindrical pressure vessels and integral heat exchangers: this represents a substantial advance in technique and is considered to improve safety. During the course of development, the reactor sizes grew from four 50 MW units at Calder Hall and Chapelcross to two 590 MW units at Wylfa, although the latter station has yet to reach full power. The steam conditions (which determine the thermal efficiency with which nuclear heat is converted into electricity) have also improved, from 320 °C and a pressure of 15·5 bars* at Calder Hall and Chapelcross to 393 °C and 97·5 bars at Oldbury. Plate 1 shows the Magnox station at Trawsfynydd in North Wales.

104. The next step taken in Britain was the development of the Advanced Gas-cooled Reactor (AGR). An aim of this development was to raise the steam temperature so as to achieve a higher thermal efficiency in the generation of electricity, similar to that of modern fossil-fuelled stations. This required the use of oxide fuel to withstand the higher core temperature and stainless steel instead of magnox for cladding. Uranium oxide is a poor conductor of heat so the fuel pins have to be much thinner in order to prevent the centres overheating; they are therefore assembled in clusters or assemblies, Plate 4. Since stainless steel tends to absorb neutrons to a greater extent than magnox, it is necessary to enrich the uranium to about 2 per cent in uranium–235.† An enormous amount of research and development was needed over many years in order to attain the new conditions, and the commercial AGR programme has been seriously delayed because of unexpected technical difficulties. The AGRs continue to use graphite as a moderator, carbon dioxide as a coolant, and to be contained in cylindrical pre-stressed concrete pressure vessels. They are of nominal output 625 or 660 MW, but we were informed that, like the Magnox reactors, they will be de-rated in service because of corrosion problems which are worse at high temperature. Figure 1 shows the locations of the AGRs and Magnox reactors, together with other nuclear facilities, in this country. There are five AGR stations, each with two reactors, which will be brought into service over the period 1976–80.

105. While the UK and France were developing gas-cooled reactors, in North America water-cooled reactors were being developed, initially for the

* 1 bar ≈ 1 atmosphere pressure.

† The enrichment varies both between different zones of the reactor core and as between the initial fuel and replacements.

Plate 1. The Magnox station at Trawsfynydd, North Wales. Each reactor generates 250 MW.

Photograph by courtesy of the Central Electricity Generating Board.

propulsion of submarines, and they have become the preferred type of reactor in almost every other country. Whereas by 1980 there will be some 21 GW* of gas-cooled reactors in operation, mostly in Britain, there will be about 205 GW of water-cooled reactors, almost all of them light water reactors (LWRs) using ordinary water both as moderator and coolant. Since light water absorbs neutrons significantly, the fuel, which is uranium oxide, has to be enriched to 3–4 per cent in uranium–235. It is clad in an alloy of zirconium called "zircaloy" which absorbs very few neutrons. There are two types of LWR. In Pressurised Water Reactors (PWRs) the water is kept under enormous pressures, typically 150 bar, and prevented from boiling; it gives up its heat to steam used in the turbine through a heat exchanger. In Boiling Water Reactors (BWRs) the water is under pressures of the order of 70 bar, and is allowed to boil and pass directly to the turbines, the steam being saturated, and separated from the water in a steam drum. Both types of LWR employ a large steel pressure vessel with very thick walls—typically 160–200 mm—and the structural integrity of such a vessel is vital to the safety of the reactor.

106. We have mentioned that deuterium is an excellent moderator because of its low absorption of neutrons. Heavy water, which occurs in ordinary (or light) water to the extent of 150 ppm and from which it can be separated by a complex (and energy-intensive) chemical process, is used as a moderator in two types of reactor that are being developed in Canada and the UK. The CANDU reactor (Canadian, Deuterium, natural Uranium) is fuelled with natural uranium oxide fuel, clad also in zircaloy, and uses heavy water both as moderator and coolant, though the moderator is at low temperature and atmospheric pressure in a calandria or tank, and the coolant is contained in a large number of slim pressure tubes, each of which contains fuel elements. Because heavy water is not only very expensive but also tends to form radioactive tritiated water (tritium, or hydrogen–3, is a 12·3 yr half-life β–emitter which is extremely difficult to contain)† the coolant is kept from boiling and used to raise steam in a heat-exchanger. In the British design, the coolant is light water which is allowed to boil, and the steam is passed direct to the turbine. This reactor is called the Steam Generating Heavy Water Reactor (SGHWR), and like the CANDU design, depends on a set of pressure tubes rather than on one massive pressure vessel as does the BWR. In 1974 the Government announced that the SGHWR was to be the main thermal reactor type for the future in the UK and it was planned to instal four 660 MW units at Sizewell and two at Torness in Scotland. It is less thermally efficient than the AGR, but the 100 MW prototype built by the AEA at Winfrith, Dorset, has proved to be very reliable.

107. The last type of thermal reactor we shall describe is called the High Temperature Gas-cooled Reactor (HTR) which uses helium instead of carbon dioxide as the coolant. The fuel is uranium oxide or carbide (carbide is a much better heat conductor and has a higher melting point) in small particles, coated with graphite moderator and fired to form ceramic pellets capable of withstanding temperatures approaching 2,000 °C. There is no metal to absorb

* 1 GW = 1,000 MW = 1 million kW. The capacity of the CEGB grid is about 58 GW.
† For the biological effects of tritium, see paragraph 55.

neutrons in the core, and so there is good neutron economy and the fuel can be taken to a very high burn-up. The main attraction of the reactor is that it can produce heat at a very high temperature, high enough to allow the helium to be passed directly through gas turbines or to provide process heat for steel making. It is also considered to be very safe. Gas-cooled reactors have an inherent advantage over liquid-cooled ones in this respect, for a sudden loss of pressure would allow the liquid to vaporise rapidly, and the cooling capacity would be reduced catastrophically. The HTR has reactivity uniformly decreasing as temperature rises, and even if there was a loss of coolant, since the fuel and moderator are intimately bonded together, there could not be a major meltdown. However only a few HTRs have been built. There are two in the United States and two in West Germany. A small one built as an international project at Winfrith, called the Dragon, was closed down in 1975 because Britain was unwilling to continue to pay its large share of the cost since it did not appear to be of use in this country in the foreseeable future and other nuclear projects were deemed of higher priority.

The breeding principle

108. We mentioned in paragraph 87 that uranium–238, which plays little part in the fission process, can be converted by neutron capture and subsequent decay into plutonium–239 which, like uranium–235, is fissile. This happens in thermal reactors, and in fact early reactors such as those at Windscale were specifically designed to make plutonium–239 for weapons purposes. (The heat was not utilised.) In power reactors the fuel is left in the reactor long enough for the plutonium produced to contribute substantially to the number of fissions, as indicated in Figure 5, but some plutonium nuclei remain in the fuel after irradiation. Plutonium–239 nuclei can capture neutrons without fissioning to become higher isotopes, only some of which are fissile, and part of the plutonium is in this form. The breeding of fissile plutonium partially counteracts the loss of uranium–235 nuclei; in principle the fuel could be reprocessed to remove the fission products, and restored to its original reactivity with less new uranium than would otherwise be needed. In fact, however, the repetition of this process would still allow only about 1 2 per cent of the uranium nuclei in the fuel to be fissioned. The return is limited by the fact that in a thermal reactor the number of fissile nuclei created is less than the number of uranium–235 nuclei destroyed; that is, the ratio of these quantities, known as the "breeding ratio" or "conversion factor" is less than one.

109. If some way could be found to increase the breeding ratio above unity, it would mean that more fissile nuclei would be created than were destroyed, and it would offer the prospect of utilising the whole of the original uranium including the very large percentage of uranium–238 which cannot be fissioned in thermal reactors. This would be a great prize. The amount of uranium available in the world is large, but the element is usually found in poor-grade ores so that much effort has to be expended in winning it. The future demand for uranium is likely to be such that it may become both expensive and scarce. The ability to breed fissionable material from uranium–238 would transform the supply position, enabling the available supplies to be used perhaps sixty

FIGURE 5

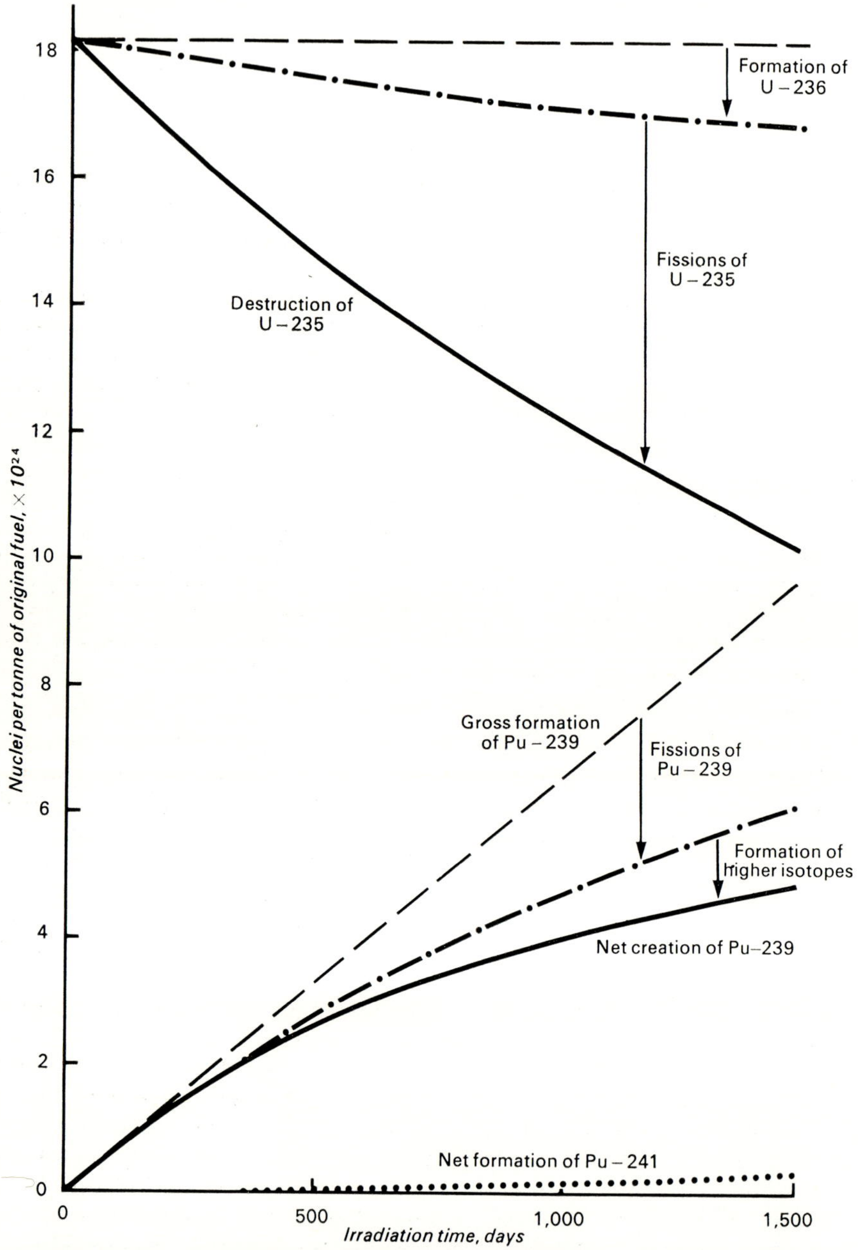

The changes in numbers of fissile nuclei in Magnox fuel during irradiation, showing the contribution made by plutonium- 239

times as effectively in producing energy. It is not surprising that major efforts are being devoted in the UK and in other countries to develop the means to achieve this.

110. In most thermal reactors, the neutron losses arising from metallic components in the core because of the use of a moderator are such that the breeding ratio is significantly less than unity. Two reactor types, the CANDU and HTR, can be designed so that the breeding ratio is very close to unity, but this does not give the optimum economic performance and some fissile material is inevitably lost when the fuel is reprocessed, so that they are not self-sustaining in fuel. A breeding ratio sufficiently above unity for this to occur can however be obtained with an unmoderated reactor, where the nuclear chain reaction is sustained by fast neutrons alone. Such a reactor is called a "fast reactor". Its design poses difficulties because fission is much less likely to be induced by a fast neutron than by one moving at thermal speed. There must be a higher proportion of fissile nuclei in the fuel and the neutron flux density (the number crossing a given area of core in unit time) must be much higher. In practice this means that the fuel must contain a substantial proportion of plutonium admixed with the uranium (from which more plutonium is bred) and that the reactor core must be very compact, resulting in a very high thermal power density. A fast reactor that can breed enough plutonium to keep itself refuelled, and have some to spare, is called a "fast breeder reactor" or FBR. We shall be mainly concerned with a particular development in which a liquid metal (normally sodium) is used as the coolant, although other systems are possible in principle, such as gas cooling.

111. Before describing the fast breeder reactor we briefly mention an alternative approach to the higher utilisation of the nuclear fuel which involves the use of thorium as well as uranium in a thermal reactor. Thorium consists mainly of thorium–232, and like uranium–238, this isotope is not fissile but it is fertile, and can capture a neutron to form thorium–233, which decays by two successive β–emissions to form uranium–233. This isotope of uranium does not occur in nature, but it is long lived (half-life 159,000 years) and fissile. A reactor with good neutron economy such as CANDU or the HTR can be fuelled with a mixture of highly-enriched uranium and thorium, and it can be designed to have a breeding ratio of nearly unity or even slightly above,* together with a high burn-up so that many of the uranium–233 fissions take place without the fuel having to be removed from the reactor and reprocessed. As a result of the utilisation of the thorium, the amount of uranium fed to a system of reactors burning both uranium and thorium would be less by a factor of about five than that needed with uranium-burning thermal reactors alone. But the technology for reprocessing and fabricating thorium fuel is hardly developed as yet, and although the thorium fuel cycle could make a substantial contribution in the longer term to the conservation of uranium, it cannot use the large stocks of uranium–238 which depend upon the plutonium-burning fast reactor for their conversion to useful energy.

* Thermal fissions of uranium–233 produce on average more neutrons than those of uranium–235 or plutonium–239.

The Liquid Metal-cooled Fast Breeder Reactor

112. The first nuclear reactor ever to produce electricity in the USA (albeit on a very small scale) was of this type, and in the UK the small Dounreay Fast Reactor went critical nearly two years before any of the CEGB's Magnox stations. But two serious accidents, involving partial fuel meltdowns, in US reactors delayed the programme there and construction work on a 350 MW prototype at Clinch River will not start before next year. The major development effort has been in Europe, which possesses little uranium ore, and thus has a greater need to use uranium economically than the USA. Medium-sized prototypes have been built in Britain, France and the Soviet Union, and one is under construction in West Germany. The Prototype Fast Reactor (PFR) at Dounreay (Figure 6, Plate 2) has a maximum rating of 250 MW, but it is envisaged that a full-scale commercial fast reactor (CFR) would be of about 1300 MW in order to drive two standard 660 MW turbo-alternators.

FIGURE 6

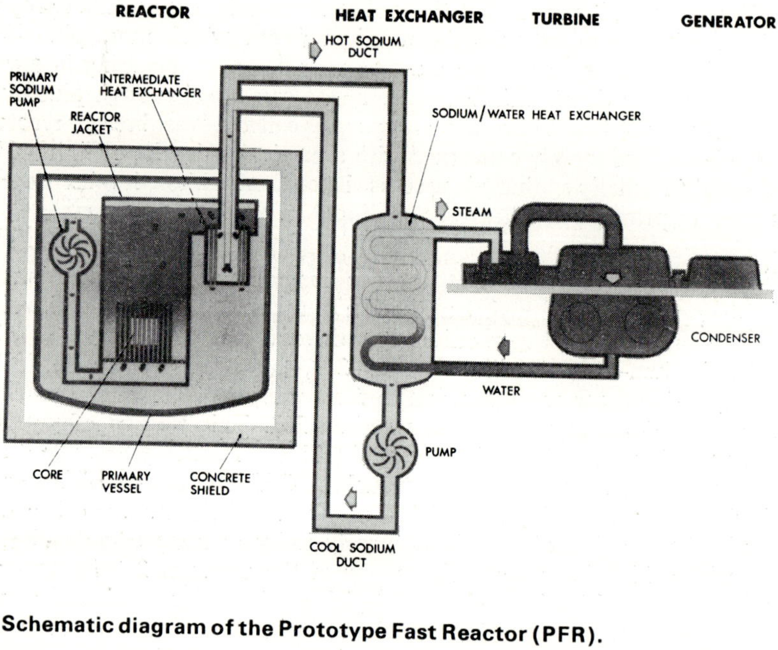

Schematic diagram of the Prototype Fast Reactor (PFR).

United Kingdom Atomic Energy Authority.

113. The fuel in the centre of the core is a mixture of uranium and plutonium oxide, typically about one-fifth plutonium, or about five tonnes in the CFR design. This provides the majority of the reactivity. Surrounding this "mixed oxide" central region are "blanket" regions in which "depleted" uranium, that is uranium from which the 235 isotope has been largely removed, absorbs neutrons and is converted to plutonium–239. (See paragraph 87.) It also acts as a neutron reflector. The fuel is removed at intervals, and the newly created

Plate 2. Dounreay Experimental Reactor Establishment, Caithness, Scotland. In the foreground is the 250 MW Prototype Fast Reactor (PFR); the sphere houses the Dounreay Fast Reactor (DFR), shortly to be de-commissioned. In the background is the fuel reprocessing works, with waste storage and disposal facilities.

Photograph by courtesy of the United Kingdom Atomic Energy Authority.

plutonium is extracted in the reprocessing plant. As its name implies, the FBR breeds somewhat more plutonium (perhaps 10–20 per cent) than is consumed in fission. Once a reactor has been built and provided with its initial inventory of plutonium (together with a margin of about 35 per cent to allow for the amount needed in other parts of the fuel cycle), it needs no further supply of uranium–235 or plutonium but will run indefinitely on the depleted uranium that is left over by the thermal reactors, and produce a net output of plutonium that can be used in other reactors. However in order to provide enough plutonium to form the initial inventory for an FBR (about 4 tonnes per GW of electrical output) it is first necessary to run a thermal reactor of equivalent output for many years. This number depends on the reactor type: for a Magnox reactor or reactors it would be about 7 years, but for an AGR, 30 years, assuming a 66 per cent capacity factor.* Thus although in the long run, FBRs offer the prospect of near independence from uranium supplies, in the short term a large FBR programme will necessitate a big thermal reactor programme and hence a big uranium demand.

114. We have noted that the core of the reactor is necessarily very compact and highly rated in terms of heat output. The power density of the core of the PFR (which is comparable to that expected in the CFR) is about 390 kW/litre, compared with 11 kW/litre in the SGHWR core and only 1 kW/litre in Magnox reactors. (By way of comparison, the power density is about 0.8kW/litre in a modern coal-burning domestic fireplace.) As a result it is difficult to arrange an adequate flow of either gas or water to remove the heat, and all the current fast reactors employ a liquid metal for this purpose. The high heat conductivity of the metal, usually liquid sodium, and the fact that the system is unpressurised are important advantages. For example, after reactor shutdown, natural convection alone will remove the fission product decay heat without the need for pumping. The use of sodium as coolant raises some problems which we examine more fully in Chapter VI. One of these is that because sodium and water react chemically if they come into contact, and because the heat exchangers have high pressure steam on one side and low pressure sodium on the other, any slight leak might lead to a vigorous reaction and dangerous consequences for the core. Consequently it is necessary to use a secondary sodium coolant circuit to convey heat from the pool of primary sodium in the core to the steam in the heat exchangers. Leaks have in fact occurred in these heat exchangers on the PFR at Dounreay and have delayed operation at high power levels by nearly two years.

115. Like the thermal reactor, the fast reactor is possible because of the accident of nature that some of the neutrons are delayed. However if there were a net addition of reactivity so large and so fast that the normal control mechanisms could not deal with it, then there exists the theoretical possibility that there could be formed a nuclear assembly that was critical on prompt fast neutrons alone. This would lead to what is technically a nuclear explosion, though the growth of the chain reaction would be slow compared with that

* Capacity factor = actual amount of electricity generated during a year/the maximum possible amount from running at the nominal rating all the time.

which occurs in a nuclear bomb and the energy release would be correspondingly less. It is not yet clear whether a nuclear explosion would vaporise fuel; it is currently assumed that this could occur and the reactor containment is designed to cope with such a contingency. However if it failed to do so, then not only iodine and caesium, but substantial quantities of non-volatile fission products such as strontium, as well as plutonium, would be released. If the reactor were in a populous area, the number of casualties could be very great. The reason why this can occur in a fast reactor, but not in a thermal one, is that in the former the fuel is not initially in its most reactive state. If all the fuel in a thermal reactor were to melt into one mass, it would be less reactive because there would be no moderator to enhance criticality. But if all the fuel in a fast reactor were to melt into a single compact mass, it would be very much more reactive.

116. The two partial meltdowns in FBRs in the USA were contained, and there was no release of radioactivity. But an uncontained meltdown could be so serious in its consequences (see Chapter VI, paragraph 303) that it is generally accepted that fast reactors cannot be major contributors to a power programme until the processes underlying the change of geometry are well understood. There is an extensive research programme in the field, but it is not yet clear whether it will prove possible so to design fast reactors as to rule out in principle the possibility of a sudden increase in nuclear power that would be so great as to rupture any feasible containment.

117. The fast breeder depends on plutonium for its fuel. This raises a number of problems, some of which we consider in detail elsewhere in the report. These include its radiotoxicity (paragraphs 66–77) and the consequent need for elaborate precautions in handling it; and the risk that it may be used for nefarious purposes, which requires that it should be carefully safeguarded. It should also be noted that plutonium is essentially a man-made element and that its supply depends on the operation of reprocessing plants. If such a plant was put out of action for a long time, as it could be, for example, by an accident leading to substantial contamination, a serious supply situation might well be created unless large stocks of plutonium existed to cover the contingency or there were alternative sources of supply.

The nuclear fuel cycle

118. In this section we describe the processes that take place outside the nuclear reactors, and their radiobiological implications. We begin with a short description of uranium mining and extraction, even though this is not conducted in the UK, as there are certain radiological problems associated with these activities. The imported uranium in the form of "yellowcake", a mixture of the two oxides of uranium, is made either into uranium metal (for Magnox reactors) or hexafluoride, UF_6, called "hex", for enrichment. The work is conducted by British Nuclear Fuels Limited (BNFL), a wholly-owned subsidiary of the UK Atomic Energy Authority (AEA) at its works at Springfields, near Preston, and Capenhurst in Cheshire, where the enrichment plant is situated. Conversion of "hex" to oxide and fabrication into fuel elements is

carried out at Springfields. After irradiation in a reactor, the fuel elements are shipped to Windscale, on the Cumbrian coast, for reprocessing to extract the unused uranium and the plutonium. The fission products are still highly radioactive and for the time being are stored at Windscale. A diagram of these activities is shown below, Figure 7.

FIGURE 7

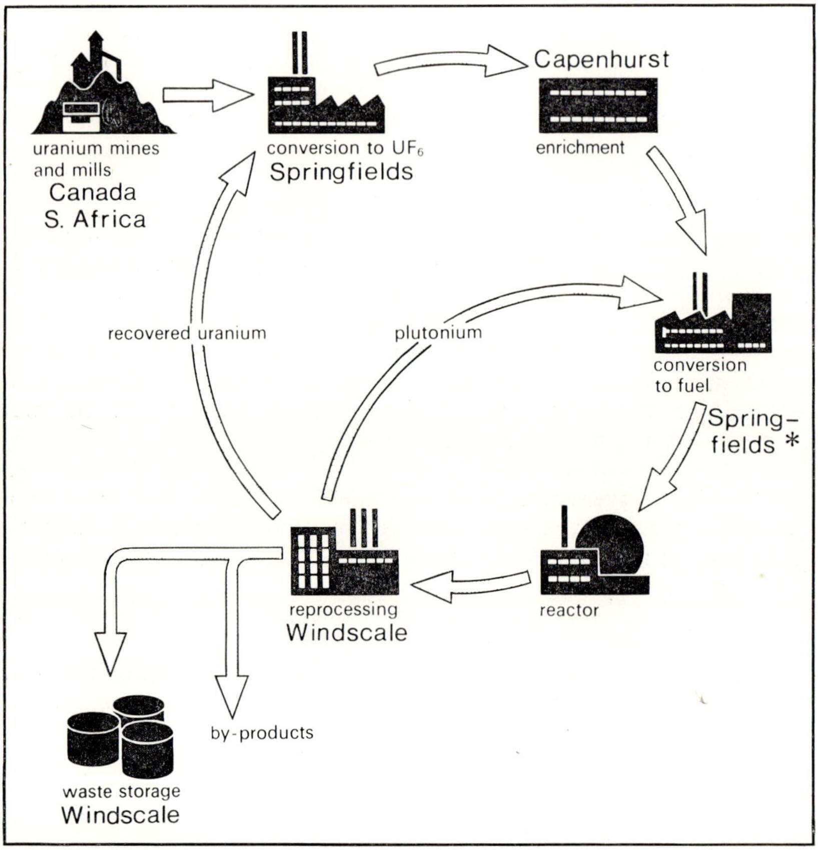

The nuclear fuel cycle.

* Manufacture of plutonium fuel is carried out at Windscale.

Walter C. Patterson: Nuclear Power (1976) p. 88

Uranium mining

119. Uranium is not a particularly scarce element in the earth's crust, but most of the known high-grade ores (containing a few percent uranium) are now worked out, and successively poorer ores are already having to be exploited. Britain's supply of uranium (a few thousand tonnes per year) comes from

Canada and South Africa; supplies are expected soon from Namibia and perhaps from Australia which has large reserves, as yet undeveloped. The ore is crushed and ground to a powder, and uranium is leached out by chemical solvents. It is concentrated as "yellowcake", and despatched in steel drums. Uranium when recently extracted is radioactive, but being extremely long-lived is only very slightly so, and the main hazard is its chemical toxicity which is similar to that of lead.

120. Because of the radioactive decay of uranium, the untreated ore also contains its decay daughters (see paragraph 34), which include the inert gas radon–222. This is present in the atmosphere in underground mines, and is inhaled by miners along with dust particles to which its own short-lived and highly active daughters adhere. In consequence the miners suffer an unusually large number of lung cancers. Radon continues to be formed by radioactive decay in the discarded tailings, and is readily released to the atmosphere because the tailings are finely divided. This is a health hazard in uranium-mining communities, as it is in parts of the United States southwest where huge piles of tailings have been created.

121. Uranium is also present in seawater (3 parts in 10^9) and the 4,000 million tonnes theoretically available dwarf the estimated deposits on land. However, although the basic process for extracting uranium from seawater is known to work and the energy cost (mainly power needed for pumping) is only one sixth that of the energy that can be obtained in a thermal reactor power station using the uranium thereby extracted([23]), much development work is still needed. Only a few areas of the world are suitable for the extraction process, as a continual supply of unprocessed water is needed, preferably warm and free from suspended matter that would clog the filters. Japan and the Bahamas are two favoured sites, but the UK is not one. The cost of uranium obtained from seawater is very uncertain, but it is likely to be several times the current price of the element (about £40 per kg.). On the other hand, the environmental impact may well be much less than that from the large-scale opencast mining of poor grade ores.

Uranium conversion and fabrication

122. At Springfields, where the yellowcake is received and stored, there are two main processes—the purification and chemical conversion of the uranium, and the manufacture of metal and oxide powder into fuel elements. The first consists of normal chemical processes and gives rise to normal chemical effluents of negligible radioactivity. The main environmental hazards arise from the need to generate and use fluorine and hydrofluoric acid, both of which are poisonous and extremely corrosive. The same is true of hex, which is a white solid at room temperature and pressure, but a gas above 56 °C.

123. We described in paragraph 92 why uranium has to be enriched in uranium–235 in order to enhance its ability to sustain a chain reaction. Since the two isotopes have identical chemical properties, the separation has to be based on their difference in mass—a little over 1 per cent. There are several

possible processes, but all depend upon the uranium being in a gaseous state, whence the need to form hex. The only one in commercial use involves diffusion through a porous membrane, which the lighter uranium–235 atoms do more readily than do those of uranium–238. The difference in rates is very slight, and a large number of stages are needed in cascade to achieve the desired enrichment. The enriched hex is obtained at the expense of a substantially greater quantity of depleted hex, containing only about 0·25 per cent of uranium–235. Throughout the plant hex is kept below atmospheric pressure to avoid outward leaks and to keep it from condensing. At each stage it is compressed and cooled, and the whole process requires a large amount of electricity. (For the AGR it is about 2½ per cent of the electricity generated by the fuel, and for the PWR, about 6 per cent.) Because both isotopes of uranium have extremely long half-lives, there is little radioactivity hazard* although precautions are taken to avoid criticality in the enriched uranium.

124. A new process is in the course of development in which the diffuser is replaced by a centrifuge (see Plate 3). This also requires many stages in order

Plate 3. Uranium enrichment by means of the centrifuge process.

Photograph by courtesy of British Nuclear Fuels Ltd.

* Highly-enriched uranium contains about 0·8 per cent of uranium–234, a third isotope of natural uranium which is shorter lived (half-life, 247,000 years) and sufficiently radioactive to require highly-enriched uranium to be handled only in glove-boxes.

to give a significant degree of separation between the isotopes, and it is more expensive to build, but its requirement for electricity is only one-tenth that of the diffusion process. Moreover the plant is more compact, easier to manage, and almost completely silent. A commercial-scale plant is now being built at Capenhurst.

125. We briefly mention two other enrichment processes that have been developed recently. In the "nozzle" process, which is available commercially, hex gas mixed with large volumes of hydrogen is allowed to expand from a nozzle in a semi-circular path: the lighter uranium–235 molecules are thrown inward and some separation results. The plant requires a lot of power but it is relatively simple and cheap to build. The laser process exploits the slight difference in vibration frequency of the two uranium isotopes. A highly-tuned powerful laser is employed to activate just the uranium–235 atoms in a mixture, which can then be separated from the uranium–238 in several different ways. This process has not yet been developed on a commercial scale, but it is potentially of great importance strategically, for a single stage of separation could theoretically produce pure uranium–235 and leave only uranium–238. This would avoid the waste of uranium–235 unavoidably left in the "tails" from a conventional plant (about one-third of the total). Moreover pure uranium–235 is weapons-grade material and so laser enrichment could be a direct route to bomb manufacture.

126. The second activity at Springfields is the manufacture of uranium fuel elements. These are normal metallurgical processes and require no special radiological precautions other than the control of criticality*. This applies only to enriched uranium fuel; for this there are strict limits on how much can be stored in one location, and great care is taken to prevent any water entering the area, as it could act as a moderator. All fuel elements have to be made with extreme precision as they are subjected to intense stresses during irradiation in a reactor, and must remain dimensionally stable, and integral so as to prevent the escape of fission products into the coolant. The fuel elements for Magnox reactors are of uranium metal and encased in magnox cans. Later designs of reactor use oxide fuel, and the uranium dioxide (which is made from the enriched hex at Springfields) is bonded and compressed into small pellets. These are heat-treated and then ground to size before being inserted in carefully machined thin stainless steel or zirconium alloy tubes. The tubes are filled with an inert gas, sealed, and assembled into composite arrays (see Plate 4). They are shipped to the reactor by ordinary transport, although subject to certain controls.

Fuel loading and discharge

127. Reactors will normally carry a supply of new fuel elements in a store and, again, care is taken that no water can enter accidentally even in the event of fire. The elements are loaded into the reactor by means of a charge machine

* Workers at Springfields are classed as "radiation workers" and suitably monitored, but the radiation doses received are minimal.

Plate 4. Inspection of uranium oxide fuel assemblies for Advanced Gas-cooled Reactors. Each of the 36 'pins' contains a large number of small sintered cylindrical pellets.

Photograph by courtesy of British Nuclear Fuels Ltd.

(see Plate 5). In the gas-cooled reactors, which have a relatively low power density, there are a large number of elements and to save time, the reactor is designed to be refuelled while running on load. (In the water-cooled reactors with higher fuel rating this is not necessary, and they are normally shut down once a year for reloading). The pressure in the charge machine is raised to match that of the coolant, the reactor core is unsealed and the spent fuel elements are removed. They are replaced by the new ones, and the reactor core is sealed once more. The irradiated fuel elements, which are intensely radioactive, slide down a heavily shielded chute into a cooling pond.

128. This pond, which is provided for all reactors (except the last Magnox station at Wylfa where there is a carbon dioxide-filled magazine) is some 6 metres deep and is fitted with racks on which the fuel elements are stored. They are both cooled and shielded by the great mass of water above them, and they glow with an eerie blue light called "Cerenkov" radiation (see Plate 6). The water is chemically treated to reduce corrosion of the cladding, and to remove fission products which may escape through failed cladding material. The shorter-lived radioisotopes decay rapidly in the cooling pond, as may be seen

Plate 5. Interior view of the AGR station at Hinkley Point, Somerset, showing the fuel charge machine positioned over the 'pile cap'. This single machine serves both of the 660 MW reactors.

Photograph by courtesy of the Central Electricity Generating Board.

from Figure 8, which shows how the radioactive decay heat in discharged fuel elements varies with time. It is normal to wait 6 months before sending them for reprocessing in order to allow the volatile iodine–131 to decay (its half-life is 8 days). By this time Magnox fuel elements would be quite cool, but CFR fuel elements would still be generating as much heat as Magnox fuel immediately on discharge, about 108kW/tonne. It will be necessary for this high thermal flux to be dissipated during the transport of CFR fuel elements to the reprocessing plant, otherwise melting could occur.

FIGURE 8

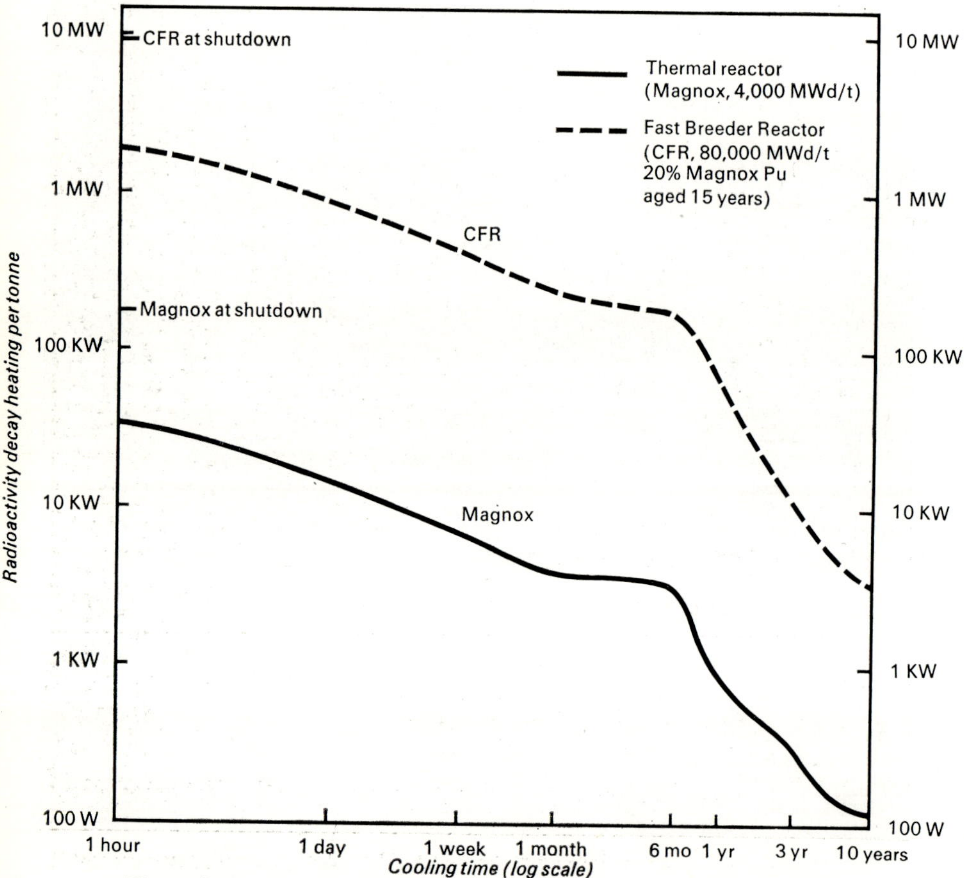

The variation in radioactive decay heating in discharged reactor fuel
Magnox 4,000 MW-day/tonne
CFR 80,000 MW-day/tonne (original enrichment, 20% Magnox plutonium, stored 15 years)

Plate 6. Cooling pond for irradiated fuel elements at the Winfrith 100 MW SGHWR. The 'Cerenkov' radiation surrounding them arises from their β-radiation.

Photograph by courtesy of the United Kingdom Atomic Energy Authority.

129. Shipment of irradiated fuel elements takes place in massive steel flasks, typically 50 tonnes, which are designed to withstand both a severe impact and a fire (Plate 7). They are loaded with fuel elements underwater, and the water within the flasks, together with the external flask fins, ensures adequate cooling during the journey by road and rail from the power station to Windscale. BNFL also do some reprocessing work for overseas customers. The flasks are transported by sea in special ships and auxiliary cooling is provided, with standby equipment to cope with any electrical failure. Before leaving the power station, the flasks are carefully decontaminated and checked for any leaks, and they are checked again on arrival at Windscale.

Plate 7. Transport of irradiated fuel by rail to Windscale, Cumbria, in fuel flasks. The square one, weighing 50 tonnes, contains 2 tonnes of Magnox fuel; the cylindrical one, weighing 75 tonnes, contains approx. 0·4 tonnes of LWR fuel.

Photograph by courtesy of Ace Film Productions.

Fuel reprocessing

130. The irradiated fuel consists of a mixture of unused uranium, some plutonium and other elements formed by transmutation, and the fission products. In the first reactors the primary object was to produce plutonium for bombs, and so methods were developed whereby it could be extracted chemically from the irradiated fuel. The extraction of plutonium still provides the main reason for reprocessing fuel: the element is valuable as a source of energy either in fast reactors or as a substitute for extra uranium–235 needed to enrich fuel in thermal

reactors. There are other reasons. One is to recover the uranium which may also be recycled if its uranium–235 content is higher than that of the "tails" (depleted uranium) from the enrichment plant, and this is nearly always so*.

131. Another reason for reprocessing irradiated nuclear fuel is an environmental one: it facilitates management of the radioactive wastes, which can be concentrated into a relatively small volume for more convenient storage. Some can be separated out for use as radioisotopes†. With Magnox fuel, reprocessing should take place within about a year of leaving the reactor because corrosion of the cladding allows fission products to leak out. But with oxide fuel, clad in stainless steel or zircaloy, both of which are highly resistant to the action of water, reprocessing can be delayed for many years. This is fortunate as at present there is no commercial-scale oxide fuel reprocessing plant in operation anywhere in the western world, and the fuel, mainly from LWRs in other countries, is accumulating in storage ponds at the power stations.

Plate 8. The Windscale works of BNFL on the Cumbrian coast. On the right are the four 50 MW reactors at Calder Hall; on the left the sphere contains the 33 MW prototype AGR. The original 'Windscale Piles' are immediately behind the tall stacks. The reprocessing plant is in the centre of the site and the route of the discharge pipelines to sea can be seen on the extreme left.

Photograph by courtesy of the United Kingdom Atomic Energy Authority.

* Recycling of uranium to thermal reactors is in practice limited by the gradual build-up of uranium–236, an absorber of thermal neutrons.

† For example, neptunium–237 is separated out and irradiated in a reactor to form plutonium–238. This has a much shorter half-life than plutonium–239 (and is correspondingly more toxic) and its radioactive decay is being used to power remotely-sited navigational buoys and heart pacemakers. The risk of accidental releases from such applications clearly requires careful consideration.

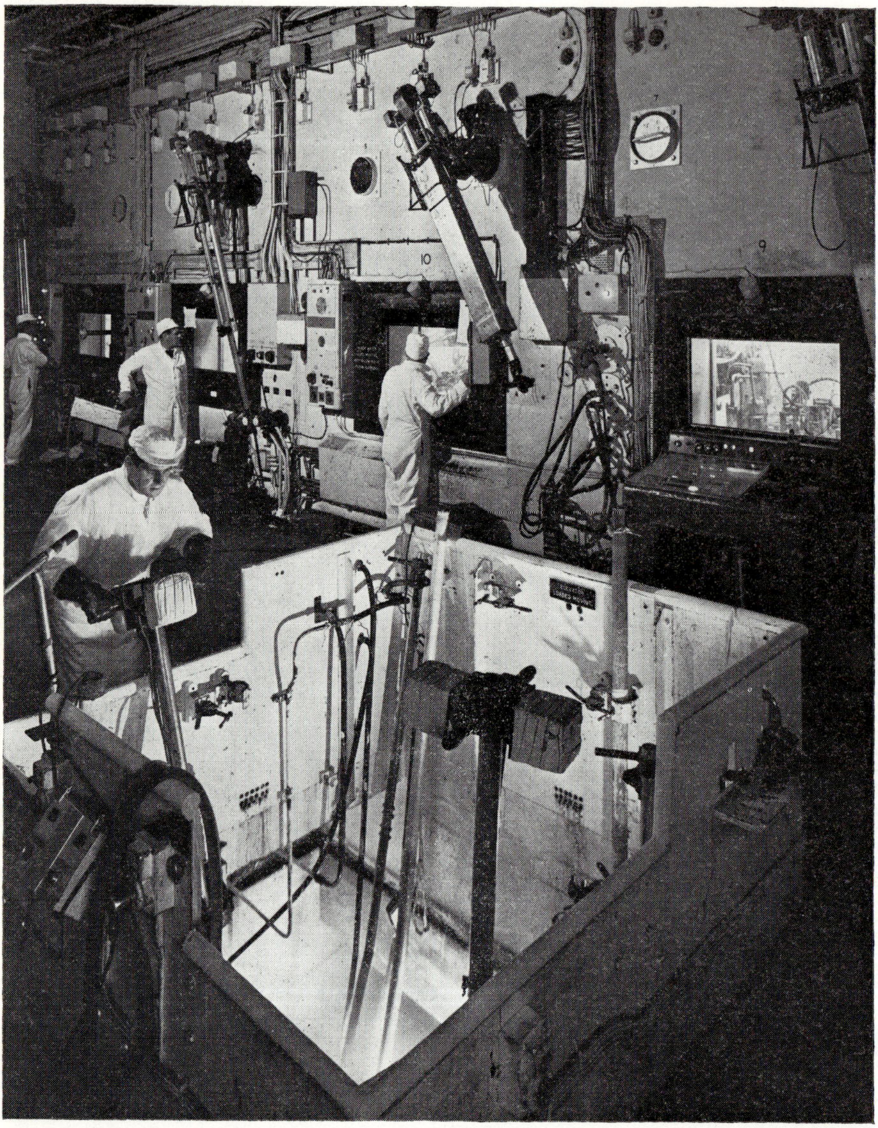

Plate 9. The reprocessing line for Magnox fuel at Windscale. The operator in the foreground is manipulating fuel elements under water. In the background are the 'caves', fitted with thick windows, in which the fuel has its cladding removed and is chopped into slices.

Photograph by courtesy of British Nuclear Fuels Ltd.

132. A reprocessing works, such as the one at Windscale (Plate 8), is in essence a straightforward chemical plant, but since the initial feedstock, all the process liquors and the products are highly radioactive all the operations have to be conducted remotely, and the design has to be such that maintenance is reduced to a minimum. Great care has to be taken to avoid criticality which is more likely to occur when fissile materials are in aqueous solution because the water acts as a moderator. As far as possible, the vessels are of such a size or shape that the geometry alone precludes the possibility of a criticality incident, but since different types of fuel are reprocessed in the same plant the concentrations of fissile material may vary considerably*. Sometimes criticality is avoided by arranging for strong neutron absorbers or "poisons" to be present—for example as a boron steel honeycomb in a cylindrical mixing vessel. This technique is particularly useful if FBR fuels, containing large amounts of fissile plutonium, are being reprocessed.

133. The fuel is taken from the storage pond and inserted in the first of a series of "caves" built of concrete several metres thick and fitted with special radiation-absorbing windows (Plate 9). The cladding is removed mechanically from Magnox fuel (Plate 10); oxide fuel elements are chopped into short lengths and dropped into acid which dissolves the fuel and most of the fission products, but does not attack the cladding "hulls" which are removed for storage in concrete silos. A few of the fission products escape to atmosphere, notably krypton–85, an inert gas with a half-life of 10·7 years. This passes through the normal air filters, and causes no problems at present. In the future, however, it may be necessary to limit the amount discharged if nuclear power increases substantially. The time required for radioactivity from the gas to decay (about a century to fall to one-thousandth of its initial level) would preclude storage in steel cylinders, but a method of retaining it has been developed. This involves absorbing the gas into active charcoal and then incorporating it in a metal matrix by spark deposition.

134. The uranium and plutonium are extracted from the acid solution with an organic solvent. This has to be stable under intense radiation, and must not react with strong nitric acid. At certain acid/solvent ratios and at elevated temperatures, a chemically explosive mixture can be created. This actually occurred at Windscale in May 1973, and some fission products escaped through the seals and into the building. The accident took place in the "head-end" plant, in which the more highly radioactive oxide fuel was being pre-processed prior to joining the main stream of Magnox fuel. The plant was thereupon closed down, but it is expected to reopen next year.

135. Plutonium is separated chemically from the uranium and is converted to the dioxide, a grey powder, for storage in stainless steel canisters. (In the USA, however, it is stored as nitrate in acid solution in carboys). The plutonium is very valuable and therefore the storage vaults have elaborate physical protection systems. It is also potentially dangerous and if not stored

* The amount of plutonium in the irradiated fuel is generally not precisely known because it depends on the irradiation history of each fuel element. This creates accountancy problems when large quantities of fuel are being handled.

Plate 10. Magnox fuel cladding being removed.

Photograph by courtesy of Ace Film Productions.

correctly could go critical. Great care is therefore taken to exclude the possibility of water entering the vaults, and a special racking system controls the geometry of the arrangement of canisters.

Radioactive wastes

136. The nuclear fuel cycle, including the operation of reactors, generates considerable quantities of radioactive waste at all stages, some of which we have referred to earlier. We discuss the wastes and their management in Chapter VIII, but we give here a brief description of some salient features in order to complete our account of the nuclear fuel cycle. Wastes can be divided for management purposes into three levels, depending on the amount and character of radioactive material present. We shall consider here mainly the "high-level" waste, containing most of the fission products, which is left after the uranium and plutonium have been extracted from the irradiated fuel. This waste must be isolated from the environment for extremely long periods and in the early years it generates so much heat that active steps must be taken to remove it in order to ensure containment of the waste. "Medium level" wastes do not pose any active threat, but must still be maintained in isolation for various lengths of time. Plutonium-contaminated waste is of this type,

and it must be contained essentially forever. Some other wastes decay quite quickly and after a few years may be mixed with "low level" wastes which are routinely discharged to the environment.

137. The high level waste is in the form of an acid solution, and consists mainly of the fission products, which are β-emitters and comprise a large number of elements, from arsenic (number 33 in the periodic table of elements) to europium (number 63). But there is also a group of radionuclides formed from uranium by successive neutron captures which are called "actinides"†. We have described how plutonium–239 is formed from uranium–238; further neutron captures lead to heavier isotopes of plutonium and, through β-decay, to two higher elements, americium and curium. Most of these actinides are α-emitters, like plutonium–239; being of shorter half-life they are extremely radio-toxic if taken into the body (curium is six times as toxic as plutonium per unit of activity) and therefore need to be very strictly controlled. We reproduce below a table of the more important isotopes.

TABLE 6

The more important isotopes of Plutonium, Americium and Curium

Isotope	*Half-life*	*Decay*	*Decay daughters*
Plutonium–238	87 yr	α	Uranium–234
Plutonium–239	24,400 yr	α	Uranium–235
Plutonium–240	6,600 yr	α, S.F.*	Uranium–236
Plutonium–241	14·3 yr	β	Americium–241
Plutonium–242	387,000 yr	α	Uranium–238
Americium–241	433 yr	α	Neptunium–237
Americium–242	16 hr	β	Curium–242
Americium–242m	152 yr	I.T.**	Americium–242→Curium 242
Americium–243	7,370 yr	α	Neptunium–239→Plutonium–239
Curium–242	163 day	α	Plutonium–238
Curium–243	32 yr	α	Plutonium–239
Curium–244	18 yr	α	Plutonium–240

* Spontaneous fission.

** Isomeric transition to the ground state of the nucleus.

138. The composition of the "high-level" waste depends on the type of reactor fuel processed. Thus, americium and curium are hardly produced at all in Magnox reactors because the uranium is relatively lightly irradiated and there is not time for the higher isotopes to be formed by successive neutron capture and β-decay. The actinide problem is more severe with fast reactors with their much more intense irradiation. In the early years after reprocessing, most of the activity arises from the fission products. The α-activity during this time is associated mainly with curium, then as curium decays, americium dominates, and finally the residual plutonium–239 with its 24,400 yr half-life is the radionuclide of greatest significance. The way in which the radioactivity generated within the high-level wastes changes with time is shown in Figures 9 and 10.

† Actinides comprise the elements following actinium in the periodic table (i.e., thorium, protactinium, uranium, neptunium, plutonium, americium, curium and heavier elements) and are named by analogy with the lanthanides, the rare-earth elements following lanthanum.

FIGURE 9

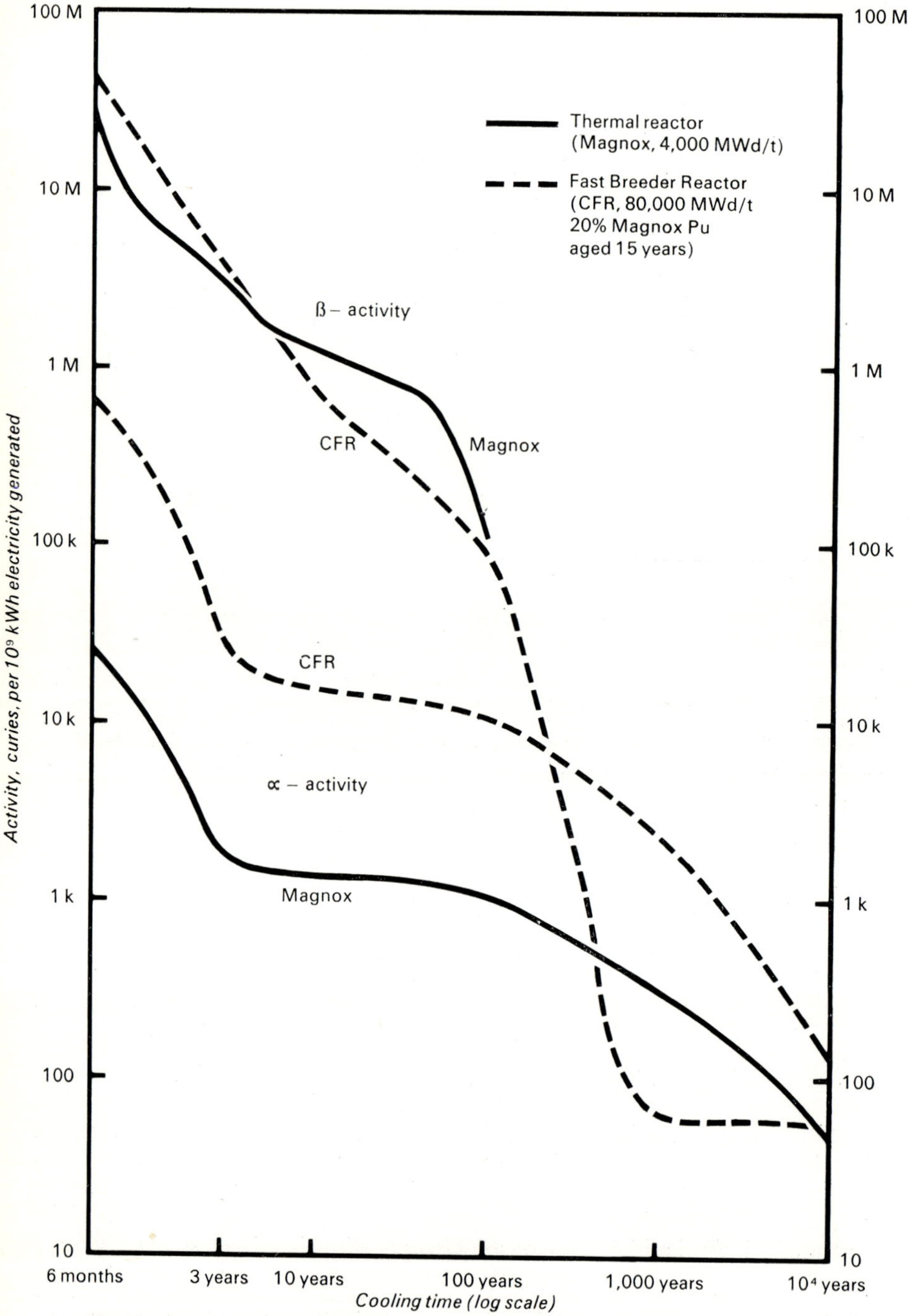

The decrease in radioactivity in high-level wastes with time for thermal and fast reactors.

Separation of 99% of uranium and plutonium at 6 months after discharge from reactor

FIGURE 10

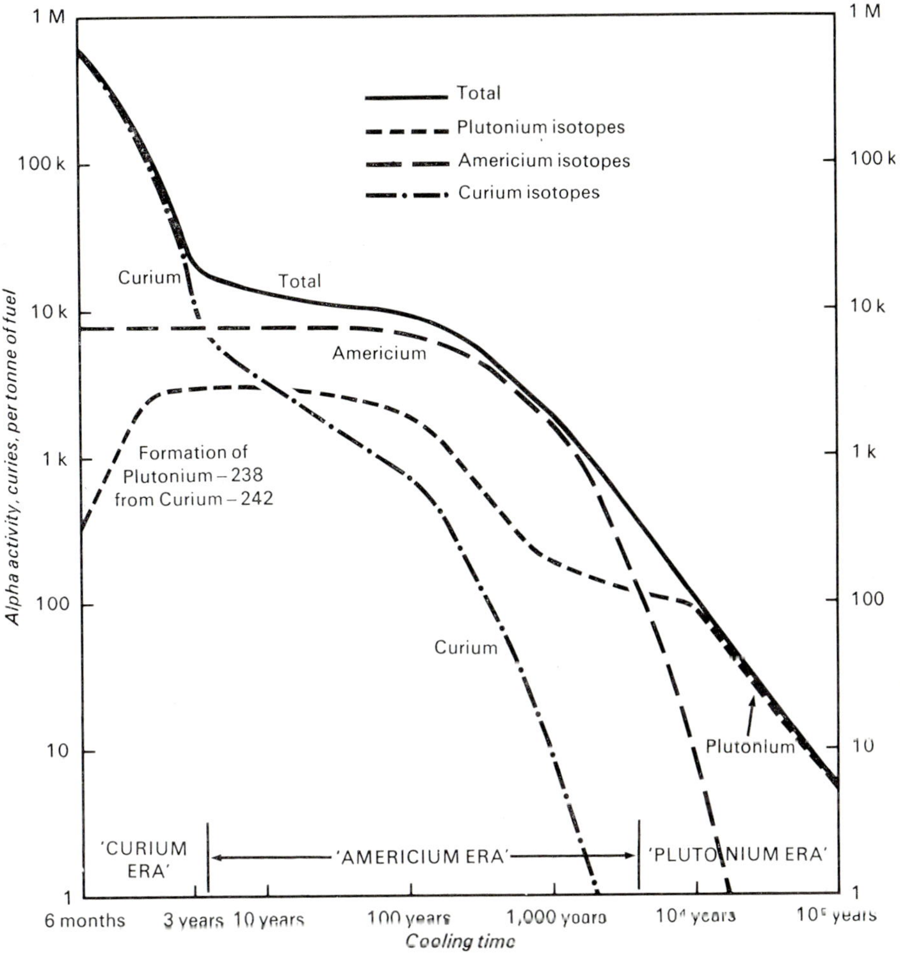

Contributions to total alpha-activity at different times in high-level wastes from a commercial fast reactor.

Separation of 99% of uranium and plutonium at 6 months after discharge from reactor. Fuel originally containing 20% Magnox plutonium, aged 15 years, irradiated to 80,000 MW-d/tonne

139. The high level waste must be securely contained against release to the environment. As an illustration of the hazard posed by high-level waste we may consider just one radioisotope, the fission product caesium–137. The throughput of this radioisotope at Windscale is at present about 18 MCi each year. The release of caesium–137 to sea is currently about 1/600 of this amount, and the resulting collective dose (i.e. the sum of all the individual whole-body doses) to the UK population, as a result of eating fish caught in the Irish Sea, is 1500 man-rem per year(6). The individual dose received by members of the

critical group is about 3 per cent of the ICRP limit. For purposes of illustration we suppose that the throughput at Windscale at some year in the future† corresponds to a nuclear power programme of ten times that at present, and that after an initial cooling period for these wastes of 20 years there were to be a failure of containment such that there leaked out into the sea just one month's throughput. We estimate that this would involve a release of 10 MCi of caesium–137, and that it would cause a population dose of about 500,000 man-rem, leading eventually to perhaps 50 deaths, from this one radioisotope alone. We do not suggest that this is at all likely to happen in view of the great precautions that are taken (see below). The example is intended simply to illustrate the need for the strictest containment of the wastes for very long periods.

140. After extraction of the uranium and plutonium, the acid solution containing the fission products and actinides is allowed to boil for a period under the influence of radioactive heating in order to reduce its volume. One tonne of Magnox fuel, which will have generated nearly 30 million kWh of electricity, produces about 40 litres of high-level concentrate: oxide fuel is expected to yield rather less. The concentrated liquid is kept in special stainless steel storage tanks. These are exceedingly elaborate, and correspondingly expensive, structures. They are of double construction to contain any possible leaks (none has occurred so far), seven independent cooling coils are provided to remove the decay heat (see Plate 11), the air above the liquid is taken off through filters to the atmosphere and, in the tanks of most recent design, provision is made for blowing air through the liquid to keep any solids in suspension rather than allowing them to settle on the tank floor. Removal of the decay heat over periods of a few hundred years is essential; failure to do this would cause the solution to boil dry and the heat generated would then disseminate volatile materials to the atmosphere and cause widespread contamination, since the filters would probably be only partially effective under such conditions.

141. The tank storage of high-level waste is considered acceptable as an interim measure but it does require continuing maintenance and an ongoing programme of tank building to cater for growth in the volume of waste and for replacement of the older tanks. At present there are twelve tanks available, eight of $70m^3$ capacity and four new ones of $150m^3$, and the total volume in storage is $610m^3$.* But it is expected that the volume of liquid will reach $1,800m^3$ by 1985 and over $6,000m^3$ by 2,000, requiring perhaps fifty large tanks, but occupying a land area of less than a hectare. In order to reduce the burden of caring for this waste there are plans to solidify it by incorporating the waste in a glass. One glass cylinder 750mm high and 500mm diameter could accommodate the wastes from the generation of 340 million kWh of electricity.**

† On the programme described by the AEA, this would be between 15 and 20 years from now.

* This was the position in May 1976. Two of the old and two of the new tanks were empty and used as spares. Four new tanks were under construction. Each costs about £4 million and takes about four years to build.

** This is about 1/640 of our present annual consumption of electricity, see Chapter IX. Cylinders of this size will shortly be produced in France, see paragraph 383 and footnote thereto.

Plate 11. The stainless steel cooling coils for a new high-level fission product storage tank being lowered into position at Windscale. There are seven independent circuits.

Photograph by courtesy of British Nuclear Fuels Ltd.

Such a process, called "vitrification", is under development but it will not be in commercial operation in the UK for another decade. The glass blocks incorporating the waste are intended to be immune to leaching by water but would still require a secure means of disposal, and a number of alternatives are under consideration such as burial in stable geological formations.

142. Intermediate level wastes are at present kept in secure storage, mainly concrete silos, at Windscale and also within nuclear reactors. They are discussed in detail in paragraphs 361–372. Low level gaseous and liquid wastes are discharged to the environment under Government authorisations. Low level solid wastes, for example used air filters and materials that become contaminated, contain relatively little activity but occupy large volumes. BNFL dispose of some very low level material at two authorised land disposal sites, and the AEA and AWRE, Aldermaston, package some of their wastes for disposal in the deep ocean. But most of the low level solid waste is awaiting a disposal means, and is simply accumulating in concrete vaults.

Plutonium as a reactor fuel

143. The final step in the nuclear fuel cycle is the recycling of plutonium to provide additional reactivity in reactor fuel, either in thermal reactors

Plate 12. The plutonium fuel element factory at Windscale where PFR fuel is made. The operator in the middle distance is working with his hands in a 'glove box'.

Photograph by courtesy of British Nuclear Fuels Ltd.

(where perhaps 1 per cent would be used) or in FBRs, where up to 30 per cent might be needed in certain parts of the core. It is used in oxide form, and because of its high radio-toxicity, it must be handled remotely in glove boxes (see Plate 12), plastic-sided boxes in which it is possible to insert one's hands through gloves permanently attached to holes in the side. (These are used where the radiological hazard is only from α- or β-emitters.) The plutonium oxide is mixed in the required quantity with uranium dioxide, and the "mixed oxide" is then formed into pellets in the same way as uranium dioxide. With a high plutonium content, particular care has to be taken to guard against accidental criticality.

144. The existing process, which has been used at Windscale to produce fuel for the Prototype Fast Reactor (PFR) at Dounreay, and also some experimental mixed-oxide fuel elements for thermal reactors, works satisfactorily, but it has the disadvantage that it generates a lot of dust, particularly from the plutonium dioxide. This not only wastes plutonium, but causes severe maintenance problems as the glove boxes have to be cleaned from time to time. Consequently a new, wet, process has been developed by the AEA for the joint precipitation of uranium and plutonium. These are formed as uniform spheres, which are washed, dried and heat treated. They are then vibro-compacted into the slender tubes used as cladding. There is no dust and the loss of plutonium is only 2 p.p.m. to the process solution as compared with 4 per cent at present. (Much of this is later recovered from the plutonium-contaminated waste treatment plant.) A plant to utilise the new process is shortly to be built at Windscale.

Thermonuclear fusion

145. We end this chapter with a brief description of fusion, which was mentioned in paragraph 84 as a possible source of energy in the future and which would have environmental advantages as compared with fission. There are several possible reactions under which light atoms can fuse to form heavier ones and release energy, but the one that is easiest to achieve is:

hydrogen–2 + hydrogen–3 → helium–4 + neutron + energy
(deuterium) (tritium)

Deuterium is a component of ordinary water (see paragraph 106) but tritium, which is radioactive, has to be bred. It is formed when energetic neutrons, such as those formed in the fusion process, bombard lithium nuclei. Thus lithium can be considered, like uranium–238, as a fertile fuel, from which tritium can be made. Very large amounts of energy are theoretically available from small quantities of materials. Thus a 1,000 MW fission reactor of the AGR type would require about 50 tonnes of enriched uranium per year, but a 1,000 MW fusion reactor would need only about 260 kg of deuterium (contained in 16,000 tonnes of water) and 440 kg of lithium.

146. Because of the strong mutual repulsion between positively charged nuclei, fusion is possible only if they have enough initial kinetic energy to overcome this. In practice, this means a temperature of the order of 100 million degrees C, and it has been established that useful energy will be produced in

quantity only if the density is also high, and the gas (which at these temperatures is completely ionised and is called a "plasma") is kept in this condition for an appreciable period of time, about a second or more. The product of particle density and confinement time must exceed a certain number: this is called Lawson's criterion.

FIGURE 11

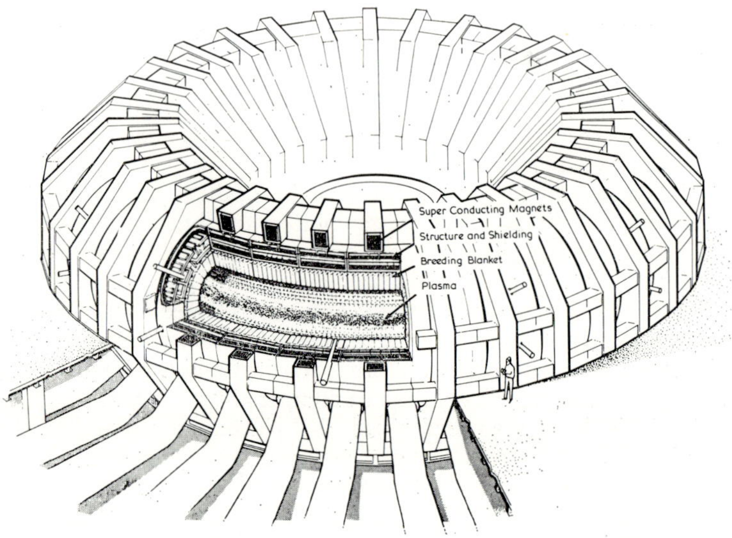

Sketch of proposed toroidal fusion power reactor (Tokamak).

United Kingdom Atomic Energy Authority.

147. The plasma has to be kept from any contact with normal materials or it would cool down. Since it is ionised and normally carries a large electric current (which heats it to the required temperature), it can be controlled by means of a magnetic field, whose lines of force act like barriers to its passage. Most experimental plasma containments are toroidal in shape; they are called Tokamaks after the Russian word for them. The plasma flows round the torus and is prevented from touching the containing wall by the action of the magnetic field produced by the surrounding coils, (Figure 11.) The major problem in development has been to maintain stability in the plasma. In present machines, temperatures of about half that required to bring about a fusion reaction, and densities and confinements of about one-tenth that needed, have been obtained, but not simultaneously. It is known that both conditions are more likely to be met with larger (and correspondingly more expensive) machines, and these are being designed and built in a number of countries. It is possible that one of these machines may achieve the required conditions, and release energy through fusion. Even so, it would still be far from a practical power station since much

more energy will be required to maintain the powerful magnetic field than is likely to be produced from fusion. It would correspond, in fission terms, to the first nuclear assembly to go critical, in 1943. It was another thirteen years before Calder Hall, the first working power reactor, was commissioned.

148. During the last dozen years, there have been notable advances in plasma physics, and there is general confidence that the problems of plasma instability can be solved. However there remain many very formidable engineering problems such as those of protecting the reactor structure from contact with the hot plasma, and from the highly energetic neutrons formed in the fusion process, of removing the heat, of circulating the lithium in which the tritium is bred, and of providing the magnetic field at acceptable cost. The development of solutions to these problems (and others) will be necessary to turn an elaborate scientific demonstration into a dependable source of power: it may well take another 20 years.

149. An alternative approach is now under active consideration in the USA and the Soviet Union. Here the confinement of the deuterium and tritium is provided not by a magnetic field but by inertial effects. A small pellet of solid deuterium and tritium is heated extremely rapidly by a high-powered laser, so that it is compressed to an enormous extent and Lawson's criterion is then fulfilled. There have been some encouraging early results, but the real problems are likely to lie in the construction and operation of lasers of a size and power far above what is now available and in the control of instabilities of the ensuing hot plasma. Although the engineering problems are likely to be just as formidable as with the magnetic confinement of plasma, they may be soluble individually, and moreover the minimum size of unit that will produce economic power may turn out to be much smaller.

150. The stated aim of fusion development in the USA is to build a demonstration fusion power station by the end of the century, but even if this is achieved it would be some decades after that before fusion could make a substantial contribution to electricity supplies. Fusion is likely to suffer from some of the disadvantages of fission, namely that it is very expensive in capital cost, that it works only in very large units (the minimum size of Tokamak may be as much as 3,000 MW), and that it is exceedingly demanding technically. On the other hand, it has some very significant advantages, particularly as regards its effects on the environment, which have already been studied in some detail([24]). The fuel requirements would be extremely small, so the fuel will be cheap and should not require the very substantial open-cast mining operations that are used for the extraction of uranium. World lithium resources are large, but there are a growing number of uses for the metal and supplies would need considerable expansion if there were to be a major world fusion power programme. In the long run, there is the possibility of developing a fusion reaction based only on two nuclei of deuterium of which the reserves in sea water are virtually inexhaustible. However the conditions under which the deuterium-deuterium reaction will take place are appreciably more difficult to realise than those for the deuterium-tritium reaction on which virtually all work has so far been concentrated. The fusion process itself as at present envisaged would

be inherently safe against runaway, so that no harm to the public would be likely to result from a loss of control. However the very powerful magnetic field used for confining the plasma would contain energy equivalent to up to 20 tonnes of TNT. In order to prevent its explosive release causing damage, the magnet would need to be surrounded by a massive containment structure. The only sources of radioactivity are the tritium used in the process, and the activation products from the intense neutron irradiation of the reactor wall. These we discuss below. There would be no fissile material present, and hence no problems of safeguards, unless (as has been suggested) the fusion neutrons were used to breed plutonium–239 from depleted uranium.

151. The inventory of tritium has been calculated as about 400 kCi per 1,000 MW reactor, and it has been suggested that routine losses could be kept within 1 part per million per day([24]). Since the Windscale works currently discharge tritium to atmosphere at the rate of 20 Ci/day (7 kCi/year) the routine discharges from even a large plant should be of little concern, and the main issue is whether a large fraction of the total inventory could be suddenly released. One possible mechanism is through a fire in the liquid lithium breeder blanket, but this could be avoided by having the lithium in the form of a fluoride salt. Even if large amounts of tritium were released, since it is of relatively low specific radiotoxicity, it would be less harmful than an equal amount of activity contained in fission products. By way of comparison, a fission reactor of the same electrical output would contain about 10,000 MCi of fission products many of which are extremely dangerous.

152. The other source of radioactivity is the reactor structure itself. Here however the designer has the possibility of selecting materials that offer a good compromise between operating temperature (which determines thermal efficiency), cost, and the half-life of the activation products. Niobium is a good refractory metal but it has a long lived neutron activation product; stainless steel is likely to be used in reactors if they are built on a large scale and it would have less long-term residual activity, between two and three orders of magnitude less than from the fission products from a fission reactor. Moreover, many of the latter are volatile and would be disseminated to atmosphere if cooling were not provided, whereas the activation products of a fusion reactor would be non-volatile solids.

Review

153. In this chapter we have described how nuclear power works, and the main operations of the nuclear fuel cycle. It is a fascinating subject for study, but it is also a very demanding technology. For the nuclear fuel cycle to work, all the parts of it must function properly. To date, the major development effort world-wide has been in the building of reactors, and the successful building and operation of so many different types is a tribute to the skill and ingenuity of the engineers and scientists who have worked together on their various problems. In the future, there will need to be comparable effort on other parts of the fuel cycle, particularly fuel reprocessing and waste management, if the whole is to operate smoothly and without danger to the public.

CHAPTER IV

MAJOR ISSUES RAISED BY NUCLEAR POWER

Introduction

154. In the previous two chapters we have described the nature and effects of radioactivity and the principles of nuclear power and the nuclear fuel cycle. Our aim has been to provide the factual basis that is required in order to appreciate the problems raised by nuclear power and its actual and possible environmental effects. We are here considering nuclear power derived from fission; the arguments about fusion power, if it is successfully developed, would be quite different. Concern about nuclear development centres on a few major issues which we consider in some detail in later chapters. We thought it would be useful at this stage, however, to survey these issues as a whole so that they may be seen in perspective and to present the underlying social and ethical questions that they raise. We begin with some preliminary comments about world energy demand and the projected scale of nuclear development, and about nuclear hazards in relation to those arising from other technological developments.

World energy demand

155. Energy is a necessity for economic growth and higher living standards and there is a steadily rising world demand. We consider energy questions in Chapter IX but we note here that great uncertainties are involved in forecasting future demand for energy. A wide range of forecasts can be prepared depending on the assumptions that are made about the rate of economic growth and population, which will interact with the assumptions made about the future prices of energy in its different forms in relation to price levels generally. Nevertheless, some general features of the energy situation are clear, in particular the threatened exhaustion of oil and natural gas (which at present supply a large proportion of energy needs, especially in industrialised countries) by about the turn of the century. On some projections of economic growth, there will be an increasing gap between world energy demand and the supplies that can be made available from fossil fuels after 1990. It appears that nuclear power is the only alternative energy source that is sufficiently developed to be capable of supplying energy on a scale commensurate with this gap, and on this timescale. Accordingly, the energy plans for the major industrialised countries envisage massive investment in nuclear generating capacity by year 2000. The long lead-times involved in the design, construction and commissioning of nuclear plant imply an early commitment to the start of these programmes and to related research and development work.

156. We should not, however, be solely concerned with the energy requirements of this and other industrialised countries. A recurrent theme of the Stockholm Environment Conference in 1972 was that poverty represented

a major form of environmental deprivation for very many people, and we should also be concerned with issues of world poverty and under-development. The question arises whether a great expansion of nuclear power in industrialised countries will help or hinder the struggle for development in the third world. It has been argued before us that for the UK and other developed countries to rely more heavily on nuclear power would release fossil fuels for developing countries. The proposition seems plausible but we are not convinced by it. The high demands of industrialised countries for fossil fuels will continue for many years, not least to sustain the economic growth which would be required to support large and costly nuclear programmes. We find it hard to see that nuclear development will release fossil fuels to an important extent unless the political will were to exist in industrialised nations to bring this about by a large international redistribution of income, and this does not appear likely. Moreover, the concentration of resources on nuclear expansion may mean that insufficient work is done on alternative sources of energy which could well be more appropriate to some of the energy requirements of developing countries. Thus, many of these countries have abundant sunshine which could be exploited as an energy source; research and development in this area would certainly be of more value to the relatively scattered rural populations, who comprise generally the poorest and most neglected section of the under-developed world.

Projected nuclear development

157. An understanding of the scale of possible future investment in nuclear power is necessary if the issues that it raises are to be seen in perspective. We consider projections of nuclear growth in more detail in Chapter IX but our present aim is to bring out the broad implications in terms of the numbers of reactors, the production of plutonium and the creation of highly radioactive wastes. For the UK we use figures relating to the programme projected by the AEA and provided in their evidence. We would emphasise that this is not a firm programme but a possible programme which is used by the AEA in their nuclear research and development planning. It is based on particular assumptions about growth in electricity demand. The details of this programme are given in Chapter IX (see Table 14, paragraph 468) where we also question its validity.

158. On this programme the total UK nuclear electrical generating capacity would increase from 5GW in 1975 to 104GW by year 2000 and 426GW by year 2030. There would be rapid growth in the use of fast reactors in the later years of the programme and, by year 2030, reactors of this type would constitute the main source, with a capacity of 370GW. Thus, by year 2030 there would be several hundred large reactors in operation and, depending on the siting policy adopted, these would be installed in perhaps 100 nuclear stations on coastal sites. There would be some growth in fossil fuel stations which would be needed to meet peaks in demand, the large base load being met by nuclear plant.

159. There would be concomitant increases in the production of plutonium and of highly radioactive wastes. Plutonium is produced in both thermal and fast reactors and would be used mainly as fuel for the latter. The amount involved in the AEA programme would increase from a little over 10 tonnes in

1975 to about 250 in year 2000 and perhaps ten times this figure by year 2030. There would need to be substantial movement of plutonium between different facilities, incorporated in new fuel elements. These movements would probably number several hundred per year by 2000 and several thousand by year 2030. The quantity of highly radioactive wastes produced would increase in rough proportion to the growth in generating capacity, though the cumulative total of radioactivity would grow much less rapidly because of the rapid decay of some of the fission products. As described in paragraph 141, the wastes up to year 2000 if stored in liquid form would occupy a volume of about 6,000 cubic metres. We have explained that this waste remains intensively radioactive for many centuries and that because of the dangerous nature and exceedingly long half-lives of some of the constituents, particularly the residual plutonium, it must not be allowed to escape to the biosphere in significant quantity.

160. We have so far dealt with possible nuclear development in the UK but it is important to appreciate the scale of projected development in world terms. Predictions of world growth in nuclear capacity are of course subject to many uncertainties and show correspondingly wide variations. Some estimates suggest(1) that by 2000 the installed electrical generating capacity may approach 3,000GW, about 30 times the figure projected by the AEA for the UK by that year, though there appear to be grounds to doubt whether such rapid growth would be possible (see paragraphs 475–476). For the USA alone it has been estimated(25) that by year 2020 there might be the equivalent of 2,000 large reactors of current type in operation, most of them breeder reactors; that there might be about 60 fuel reprocessing and fabrication plants in operation; that plutonium generation might exceed 30,000 tonnes in total; and that, assuming a dispersed industry, about 100,000 shipments of plutonium might take place annually. Such proliferation of reactors, nuclear plants and shipments of nuclear materials throughout the world will certainly greatly increase the probability of accidents and the opportunities for malevolence.

161. The plutonium already created amounts to a little over 10 tonnes in the UK and perhaps an order of magnitude more in the whole world. (Of course, these figures do not include the quantities involved in weapons programmes, about which we have no knowledge.) Nuclear development on the scale indicated above, however, would create a new situation in which plutonium, notwithstanding its dangers, would be in widespread use as a staple commodity of energy supply; it would lead to what has been called *the plutonium economy*.

Nuclear power and other technologies

162. Before discussing the hazards of nuclear power we would emphasise that in many respects these are not unique. Perhaps because of the deep-seated anxiety that is felt about radioactivity, there is a tendency to dramatise the risks in ways which may convey quite misleading impressions to people who have no basic understanding of the subject. It is said, for example, that a piece of plutonium the size of an orange contains enough of the substance to kill everyone on the earth. So it does, but it is impossible that it could be so distributed as to have this effect. A very similar statement might be made about

other substances that are commonly produced and used in industrial societies. For example, chlorine is a very basic material in the chemical industry and is made in the UK alone to the extent of about 1 million tonnes per year. Yet a mere 10mg would be lethal if inhaled; little more than a two-millionth part of our annual production would, in theory, suffice to kill the entire population. It is important to see nuclear hazards in perspective and we have tried to do this throughout our study.

163. We realise, however, that to say that other hazards exist is not to say that they, or those from nuclear power, should be accepted. With increasing environmental awareness there is a growing disposition to question the benefits of technological developments and to be concerned with their possible consequences. A recent report([26]) has stressed the need for the most searching examination of the implications of major technological developments, at the stage when these are conceived and before any commitment to them is made. Imaginative assessment of possible long-term effects, environmental and social, might disclose implications which, though necessarily speculative, might entirely change our attitude. The consequences of nuclear development on a massive scale certainly merit such examination. The report also refers to a general problem which we have ourselves encountered in the nuclear field during our study; that is, the difficulty of obtaining independent but expert advice on technical questions, since the acknowledged experts are often themselves involved in the related developments.

164. With these preliminary remarks we now turn to the arguments advanced against nuclear power. We consider them under the following headings:—

(*a*) the proliferation of nuclear weapon capability;

(*b*) reactor safety;

(*c*) radioactive waste;

(*d*) the diversion of plutonium;

(*e*) the uncertainty of radiological standards.

The proliferation of nuclear weapons capability

165. Nothing would be more disruptive to the environment than nuclear warfare and we feel bound to comment on the spread of civil nuclear power in relation to the proliferation of nuclear weapons. We list in Table 7 by quinquennia those countries that have or are expected to have commercial nuclear reactors in operation. The Table also shows the status of the nations concerned with respect to the Non-Proliferation Treaty (NPT). In ratifying this Treaty, nuclear weapon states agree not to transfer nuclear weapons or manufacturing technology to non-nuclear weapons states, and non-nuclear weapons states equally undertake not to manufacture or acquire these weapons, and to accept a system of controls and safeguards designed and operated by the International Atomic Energy Agency (IAEA) to prevent diversion to non-peaceful use of fissile material produced and used in their civil nuclear programmes. Ratification of the Treaty indicates acceptance of its principles by the states concerned.

Following ratification each state has to negotiate with the IAEA to establish the details of the safeguards arrangements as they will apply to its civil nuclear installations.

TABLE 7

Countries operating or expecting to operate commercial nuclear reactors above 100 MW, with their status under The Nuclear Non-Proliferation Treaty (NPT)

Quinquennium	*Cumulative total*	*Countries (s=signed, r=ratified)*		
1955–59	2	UK (r)	USA (r)	
1960–64	5	USSR (r)	Italy (r)	France (–)
1965–69	11	Japan (r)	W. Germany (r)	Canada (r)
		India (–)	Spain (r)	Switzerland (r)
1970–74	19	Pakistan (–)	Czechoslovakia (r)	Netherlands (r)
		Argentina (–)	E. Germany (r)	Belgium (r)
		Sweden (r)	Bulgaria (r)	
1975–79	25	Korea (r)	Taiwan (r)	Finland (r)
		Austria (r)	Mexico (r)	Brazil (–)
1980–84	32	Yugoslavia (r)	Iran (r)	Hungary (r)
		South Africa (–)	Philippines (r)	Egypt (s)
		Israel (–)		

166. The Table indicates that about fourteen countries per decade are in the process of acquiring nuclear reactors, and with them the material with which nuclear weapons may be made. Although the plutonium cannot be extracted without the use of an elaborate reprocessing plant, there are plans to install such a plant in a number of these countries. In particular, we are greatly concerned about the intended sale of these plants to countries that have not signed or ratified the NPT, such as Brazil and Pakistan. Even if irradiated fuel is sent abroad for reprocessing, the countries concerned are likely to require that the plutonium be shipped back in pure form. It will, of course, be subject to safeguards, such as those prescribed by the NPT. But a number of the nations are not parties to this treaty, and even those that are can terminate their commitments on three months' notice whenever unforeseen events have placed the vital interests of the state at risk. In the last resort, international inspectors can operate only by the consent of individual governments, and this might be withheld on short notice. There is nothing to stop the manufacture of all the non-nuclear parts of a nuclear bomb and of facilities that would permit returned plutonium to be fabricated into a suitable chemical and physical form when required. A possible check would be for all separated plutonium destined for non-nuclear weapon states to be returned only in the form of fuel elements for a reactor, these being made under international control, and perhaps briefly irradiated in a reactor to make the plutonium inaccessible. There is, however, little prospect of acceptance of such a system by states that are not parties to the NPT and it could be circumvented by those who were prepared and able to acquire and operate fuel reprocessing facilities.

167. We welcome the recent "secret" meetings of nuclear-exporting countries to discuss the proliferation problem and the safeguards that should be applied to the export of nuclear materials and facilities. Agreement on these matters would certainly reduce the risks and is highly desirable. Nevertheless, we do not

think it would alter our conclusion that the spread of nuclear power will inevitably facilitate the spread of the ability to make nuclear weapons and, we fear, the construction of these weapons. In reality, total agreement on a comprehensive international control system for the products of civilian nuclear power that are relevant to the construction of nuclear weapons would be possible only in a climate of general disarmament, and the prospects for this are receding rather than improving. It has been argued that the possession of these weapons by the USA and the USSR has been a powerful force for mutual toleration, but however true this is it would be folly to suppose that proliferation would necessarily lead to a similar balance and restraint in relations between other nations. Indeed, we see no reason to trust in the stability of any nation of any political persuasion for centuries ahead. The proliferation problem is very serious and it will not go away by refusing to acknowledge it.

Reactor safety

168. We have explained that the materials within an operating reactor contain a vast quantity of radioactivity, and security of its containment under fault conditions is thus a major issue. Proponents of nuclear power point to the fact that in the design and construction of reactors very great efforts are devoted to the analysis of possible failure modes and to the incorporation of safety precautions to guard against their occurrence. Still, as described in paragraphs 100–101, it is possible to postulate failure mechanisms, however unlikely, which might result in the accidental release to the atmosphere of some fraction of the gaseous and more volatile fission products contained in a thermal reactor core. As explained in paragraph 115, failure mechanisms can be envisaged for a fast reactor whose consequences would be intrinsically much more serious. A major release of radioactivity would have three types of effect. Direct radiation from the airborne cloud could cause acute medical effects (including possibly some deaths) among people fairly near to the site; there would be long-term medical effects, principally the induction of cancers, which would develop over a period of 10–20 years after the event; and there would be contamination of ground and property by the deposition of radioactive material which might require the evacuation of some areas for months, or even years, depending on the levels of radioactivity and the feasibility of decontamination.

169. We discuss reactor safety and the consequences of possible accidents in more detail in Chapter VI and we restrict ourselves here to some general comments on the matter. Much work has been done to assess the reactor accidents that could occur and their possible consequences. We accept in this matter the expert view that has been put to us that in highly complex technology of this kind it is not feasible to specify a worst accident whose environmental effects would be so small as to be accepted without concern, and to ensure by design that no accident of greater severity could occur. There is a range of accidents and consequences spanning many orders of severity with varying probabilities of occurrence. The practical aim of design can only be to ensure that the probability of an accident is sufficiently small in relation to its possible consequences.

170. The risk of death or injury through accident is a condition of living and many human activities imply a judgment that the benefits outweigh the related risks. The following Table gives some examples:—

TABLE 8

Probability of death for an individual per year of exposure (Orders of magnitude only)

Risk	*Activity*
1/400	Smoking (ten cigarettes a day)
1/2,000	All accidents
1/8,000	Traffic accidents
1/20,000	Leukaemia from natural causes
1/30,000	Work in industry
1/30,000	Drowning
1/100,000	Poisoning
1/500,000	Natural disasters
1/1,000,000	Rock climbing for 90 seconds* Driving 50 miles*
1/2,000,000	Being struck by lightning

* These risks are conveniently expressed in the form indicated rather than in terms of a year of exposure.

171. The Table is intended simply to give an appreciation of the scale of risks associated with a range of activities. Some of these risks, such as those arising from smoking or from rock climbing, are accepted voluntarily by individuals. Others are concomitants of our everyday lives, such as risks at work or from aircraft or car accidents. Another category of risks we attribute to acts of God. Generally, risks greater than about 1 in 1,000 per year are considered unacceptable. With risks of the order of 1 in 10,000 per year public money will be spent to try to eliminate the causes and to mitigate their effects. Risks below 1 in 100,000 are considered as individual risks and warnings may be given. Risks below 1 in 1 million per year are generally accepted without concern.

172. The risks we have discussed above apply to individuals. It is necessary to distinguish, however, between the risks of events that affect people individually or in small numbers, and which are fairly readily accepted as inevitable, and those that cause death or injury to many people at one time and which have, in proportion to the numbers affected, a much greater impact in creating public concern. Nuclear accidents would pose a threat to communities and what needs to be ensured is that over the whole spectrum of possible accidents and consequences, in terms of the numbers of people who might be killed or injured, the risks are sufficiently small as to be acceptable in relation to the benefits of nuclear power. In judging this acceptability we would have in mind the risk from other events, man-made or natural in origin, whose effects could be on a similar scale. We note here that the nuclear industry is not alone in presenting the risk of a major accident that could lead to the deaths of many people, though it appears unique in the persistence of the contamination that such an accident would cause. The chemical process industry may be instanced in particular as one where conceivable accidents could cause immense damage. The explosion at Flixborough brought a new awareness of these risks and is resulting in the more general application to other industries of techniques of safety analysis which have long been a feature of the nuclear industry.

173. Risks of the kind illustrated in Table 8 are known from extensive data on past occurrences. In contrast the risk of a serious reactor accident can only be estimated theoretically by an analysis of possible failure mechanisms and their probabilities of occurrence. We describe the techniques used in such analysis in Chapter VI and here simply record that we have received expert evidence that it is an attainable design objective to reduce the probability of a reactor failure leading to a substantial release of radioactivity to the range 10^{-5} to 10^{-6}/year. Taking account of the fact that the area affected by a release would depend on wind direction at the time, these figures would imply that the risk of death to an individual living near the reactor site would certainly be less than 10^{-6} a year, a level of risk which is generally regarded as negligible. As we have noted above, however, we should be concerned not so much with the individual risk as with the total effect of possible accidents. We reserve consideration of this aspect to Chapter VI but in the following paragraphs we discuss a general argument that is advanced concerning the reliability with which the risk of reactor failure can be predicted.

174. One argument advanced by opponents of nuclear energy concerns man's inability to foresee reliably all the failure mechanisms that could lead to serious reactor accidents and to introduce appropriate safeguards. Reactor technology is extremely complex; man is fallible. It is impossible to achieve absolute certainty that all engineering failures or human errors that could result in disaster have been identified through safety analysis, or that the risks have been reduced to negligible levels by design. On this viewpoint, a theoretical assessment of risks is one thing, but what is achieved in practice quite another. To some critics, therefore, calculations of the risk of disastrous failure in a complex technology, though carried out with undoubted thoroughness and integrity, are essentially irrelevant. They fear that the real risk will be dictated by human fallibility; that it will arise through circumstances which have not been foreseen and that it will become known only through experience of disasters which should not be countenanced. Such critics point to the fact that it is inherent in organisations responsible for technological development to believe in their ability to devise adequate safeguards, and that by their nature they are unlikely to address themselves to the question whether the degree of certainty about performance which is called for by the scale of the hazard is attainable.

175. There are thus conflicting views on reactor safety. There are those who believe that technical solutions can be found which will ensure that the risks of serious accidents are reduced to acceptable and negligible levels. There are others who believe that the potential hazards are so great, and the possibility of human error in devising safeguards to cover every contingency so inescapable, that an acceptable level of safety cannot be guaranteed. It is hard to see that these views can be reconciled *a priori*. The contacts we have had with the nuclear industry during our study leave us in no doubt that the most diligent attention is given to safety in the design, construction and operation of reactors. It is, however, a fact of everyday experience that all eventualities cannot be foreseen even when the most stringent precautions are taken. A commonly quoted incident in this context is the fire in a lunar module which killed three American astronauts in 1966 and which occurred because of the high

inflammability of materials in a pressurised oxygen atmosphere. This had not been appreciated in spite of the immense resources devoted to safety in the project. A more directly relevant incident is one that occurred in 1975 at Brown's Ferry nuclear power station in the USA, involving fire in the cable room beneath the reactor control room. The fire was started by the flame of a candle which was being used by a workman to detect air leaks through cable openings. The emergency core cooling systems on one reactor were put out of action by the fire and very serious consequences were rather narrowly averted. The risk of fire in the inflammable cables was realised by some of the staff and had been brought to the attention of the management, but no action had been taken. No doubt the risk of fire from any cause should have been foreseen during design and no doubt it will be covered in future, but the question arises of what other unforeseen risks may exist. Certainly it is clear that the unexpected hazards are not necessarily only the small ones.

176. Nevertheless, a view must be taken on this. The human fallibility argument is one that, pressed too far, would set an arbitrary and unduly restrictive limit on technological development. It is imperative that there should continue to be the most rigorous application of safety techniques in the design and operation of reactors. Given the emphasis on safety in the nuclear field, the measured and cautious approach to development and the extent of the precautions taken in design to limit the consequences of possible failure mechanisms, we do not doubt that the risk of serious accident in any single reactor is extremely small. The hazards posed by reactor accidents certainly do not appear to us to be unique in scale and of such a kind as to suggest that nuclear power should be abandoned for this reason alone, though the possible effects of conceivable accidents and the uncertainties involved in assessing the risks are clearly factors which should be weighed in decisions on nuclear power, as they should be for any other technological development. We note here that as the number of reactors installed throughout the world increases, so will the risk. A major accident leading, say, to several hundred deaths and hundreds of millions of pounds of property damage, though a national disaster, could certainly be sustained, but it might have devastating consequences for the nuclear industry. Such accidents appear more likely to occur in less developed countries having little technological infrastructure or tradition, but the repercussions in the UK and elsewhere could still be considerable. This is an aspect that needs to be borne in mind in thinking on nuclear strategy.

177. Damage to reactors (or to other nuclear facilities) that might result in the release of radioactivity could be caused by enemy action in the event of war, by terrorist activity or sabotage, or by natural events such as earthquakes. Installations providing vital energy supplies could well be prime targets in war. The safeguards introduced against severe nuclear accidents, including the massive structures that are used to contain highly radioactive materials, would certainly provide a measure of protection. Nevertheless, it is possible to postulate events of this kind whose consequences would be much more serious than those of a conceivable accident, and clearly the estimates of probability relating to accidental releases would no longer be entirely relevant. We consider this matter further in Chapter VII, but we note here that vulnerability to attack is one argument

against the concentration of reactors and other nuclear installations in "nuclear parks" (see paragraph 184); it is also a reason why the underground siting of nuclear installations is advocated by some experts.

Radioactive waste

178. The highly active waste which arises from fuel reprocessing contains the fission products and actinides created by nuclear fission. The waste must be isolated until the various radioisotopes have decayed to insignificant levels. For the fission products this requires a period of perhaps a thousand years, but some of the actinides have half-lives of thousands or tens of thousands of years or longer. There are theoretical possibilities for transforming these elements into less long-lived forms (see paragraphs 384–388) but the process would pose considerable technical difficulties and operational hazards and appears unlikely to offer a solution in the foreseeable future. We must assume that these wastes will remain dangerous, and will need to be isolated from the biosphere, for hundreds of thousands of years. In considering arrangements for dealing safely with such wastes man is faced with time scales that transcend his experience.

179. The creation of wastes which will need to be contained for such periods of time, and hence of a legacy of risk and responsibility to our remote descendants, is a matter of great concern to many people. We think, however, that some continuity must be assumed in human affairs and institutions and in the ability of future generations to maintain the necessary containment. To suppose otherwise is to postulate breakdown in our society and our institutions, and thus conditions in which the hazards from radioactive wastes could well appear to be one of the less dire problems facing mankind.

180. Moreover, concern about the possible effects of our present actions in the remote future should not be limited to radioactivity. The widespread discharge of some pollutants to the environment, for example, of heavy metals to the oceans, though adequately controlled on the basis of present understanding, could prove to be a legacy of risk to our remote descendants and create problems just as serious as those of preserving the localised containment of radioactive wastes. There could be unexpected pathways back to man or effects on the natural world in the remote future; unlike radioisotopes, heavy metals do not decay but remain toxic forever. We do not imagine that anyone who was aware of the sufferings of the victims of mercury poisoning at Minamata would be particularly impressed with claims made for the uniqueness of pollution by radioactivity.

181. We have, however, no intention of minimising the considerable problems posed by the need to ensure the containment of highly radioactive wastes. There is general agreement that the present practice of storing these wastes in tanks in liquid form and in essentially accessible facilities, although safe enough for the present, is unacceptable as a long term solution. Processes are under development to convert the wastes to solid and virtually insoluble form, so facilitating safe storage. But the problem of isolation for immense

periods of time still remains. There are various proposals for the permanent storage of the solidified wastes under the ocean bed or in deep geological formations where they will be virtually inaccessible. Much research is needed (though still on a very modest scale compared with that on reactors) to establish the safety and the feasibility of such methods, and we give our views on what should be done in Chapter VIII. We are confident that an acceptable solution will be found and we attach great importance to the search; for we are agreed that it would be irresponsible and morally wrong to commit future generations to the consequences of fission power on a massive scale unless it has been demonstrated beyond reasonable doubt that at least one method exists for the safe isolation of these wastes for the indefinite future. The use of this method might appear unjustifiably expensive and alternative approaches might be adopted in the event. But it would, and should, be available as a fall-back position if circumstances demanded it.

The diversion of plutonium

182. Many people who are concerned about the implications of nuclear power development see a great threat in the potential attraction of plutonium to terrorist or criminal organisations. One reason for theft would be the value of the element: if plutonium-fuelled reactors become common then it will be traded internationally. One kg of plutonium can produce as much energy in a power station as 1,700 tonnes of oil, currently worth about £80,000. The other reason would be because of its potential use in a terrorist weapon, which would have enormous psychological impact. It could be disseminated into the atmosphere with conventional explosives, when it would pose not only acute and long-term radiological hazards to those who inhaled the airborne particles, but would contaminate large areas of land. Decontamination would be very costly; an estimate of the cost, given to us in evidence by the NII, is several hundred thousand pounds per gram of dispersed plutonium. There is also the risk of the construction of a nuclear bomb by an illicit group. We consider these matters further in Chapter VII, but our view is that it cannot now be regarded as beyond the capabilities of a well-organised and determined group to construct a crude but very effective weapon. In sum, plutonium appears to offer unique and terrifying potential for threat and blackmail against society.

183. The problems of safeguarding society against these hazards could become formidable in a "plutonium economy". There are particular risks during transport of the element between nuclear installations, although techniques could be adopted to make access to the plutonium both dangerous and difficult. There is also, however, the risk of theft of plutonium by direct action at installations where it is stored or by people working in the industry. Of course, many measures are taken to prevent this but it cannot be entirely ruled out. In order to counteract these risks, some people foresee the need for the creation of special security organisations which, because of the vast potential consequences of plutonium loss, would need to exercise unprecedented thoroughness and vigilance to safeguard the material while significant quantities remained on the earth in accessible form.

184. A belief that the necessary vigilance and continuity could not be adequately guaranteed in any normal organisation led Alvin Weinberg([27]) to postulate a "nuclear priesthood"; this would be a dedicated, self-perpetuating body of people forming a technological *élite* which would be entrusted through the generations with the task of safeguarding society from the hazards of nuclear power. The idea of such a "priesthood" may seem untenable, but it is an indication of the extent of the anxiety felt by some responsible people about the hazards. Concern about security also led Weinberg to suggest that "nuclear parks" should be established. By this is meant the siting of reactors and related fuel fabrication and processing facilities in large, self-contained nuclear complexes in order to facilitate security arrangements, particularly by eliminating the need to transport plutonium.

185. Many people are concerned about the implications for society of the security arrangements that might become necessary in a plutonium economy. An effective security organisation could not be merely passive, simply reacting to events. It would need to have an active role (as was recommended for the USA in the Rosenbaum report([28]); that is, to infiltrate potentially dangerous organisations, monitor the activities of nuclear employees and members of the public and, generally, carry out clandestine operations. It would also need to have powers of search and powers to clear whole areas in an emergency. Such operations might need to be conducted on a scale greatly exceeding what would otherwise be required on grounds of national security in democratic countries. The fear is expressed that adequate security against nuclear threats will be obtained only at the price of gradual but inexorable infringements of personal freedom.

186. We are sufficiently persuaded by the dangers of a plutonium economy that we regard this as a central issue in the debate over the future of nuclear power. We believe that we should not rely for something as basic as energy on a process that produces such hazardous substances as plutonium unless we are convinced that there is no reasonably certain economic alternative. In forming a view on the future need for large-scale dependence on plutonium, therefore, we should consider the possibility and implications of alternative options for energy supply, and we discuss this aspect further in Chapter IX. We should also be aware of the advantages that nuclear power offers and which we refer to at the end of this chapter; though we do not think that these outweigh the risks we have described, they are real and should not be minimised.

The uncertainty of radiological standards

187. Some people contend that scientific knowledge is at present inadequate to define the risks from exposure to radioactivity, and dispute whether the standards chosen for permissible exposures are sufficiently safe. We commented on one such issue, relating to the effects of plutonium particles, in paragraphs 68–71. Although there may indeed be uncertainty over some of the standards, there is nevertheless an impressive body of knowledge on the subject that has been built up over nearly a century. It is difficult to think of any other form of pollution that has had its effects on man so well examined as has

ionising radiation. The ICRP, whose expert publications are universally recognised as being authoritative, is supported by many other international scientific review bodies. Although it would be foolish to suppose that all the standards are so well established that further research work will not require their amendment, there does appear to be more agreement over what can be tolerated than there is for most other pollutants. We remarked in our Fourth Report (paragraphs 144, 147–8) on the situation with lead, mercury and cadmium. The effects of quantities too small to cause symptoms of poisoning are not known with precision, and indeed are widely disputed by experts. And the technique of making a special study of the doses received by a critical group, long common in the control of radioactivity, seems for other pollutants to be still in its infancy.

188. For radioactivity as for many other pollutants we must balance its effects, and the uncertainties that attach to them, against the related benefits. So far as assessing the effects is concerned, we can do no better than require that the agencies concerned are expert, open and independent, and such as to ensure that necessary research is undertaken and that assessments are appropriately revised in the light of new knowledge. The requirement reflects on organisational arrangements which we consider in the next chapter.

The case for nuclear power

189. We have presented in this chapter the principal arguments that are advanced against nuclear power. They have an environmental and ethical basis and it is proper that we should consider them. We are conscious, however, that arguments can also be advanced on similar grounds in favour of nuclear power and that our discussion would be one-sided if we were not to recognise them.

190. Potentially, nuclear power represents a vast additional energy source for mankind which could make a major contribution to raising living standards. We consider energy questions in more detail in Chapter IX but note here that if world uranium resources as currently estimated could be completely fissioned through the use of fast reactors the amount of electrical energy produced would be many times that available from all known fossil fuel reserves. Many experts in energy matters see nuclear power as the only prospect for meeting future world energy needs and would argue that the world faces a stark choice between a nuclear future and economic stagnation and decline. If reactors and the rest of the nuclear fuel cycle work as intended, that is, if we set aside the risks arising from possible accidents or malicious acts, then nuclear power is in many ways an attractice source of electricity, offering a number of advantages over alternative approaches that might be available, which would almost certainly imply the much greater use of coal. The progressive introduction of nuclear power stations would conserve fossil fuel reserves for uses such as feedstock for the chemical industry and as portable fuel for transport.

191. The routine environmental effects of nuclear power may well be much less damaging than those of fossil fuelled power stations. Without nuclear

power the amount of sulphur dioxide(SO_2) produced would probably have to be much greater. The discharges of radioactivity from nuclear stations, both to air and water, are radiologically insignificant and there are no discharges of offensive gases or particulates. Although discharges of the latter from fossil fuel stations have been much reduced, they still pose a number of air pollution problems as recent reports of the Chief Alkali Inspector testify. Nuclear stations in the UK are nearly all on coastal sites but they need not be unsightly, as Plate 1 shows. Routine discharges from fuel reprocessing works pose more difficult problems, especially that of the safe containment of highly radioactive wastes. As we have said, however, we believe that satisfactory solutions can be found to these problems.

192. One inevitable consequence of burning fossil fuels, particularly coal, is the production of large volumes of carbon dioxide(CO_2). There has long been concern that CO_2 might eventually reach such a concentration in the atmosphere as to imperil the world's climate, or substantially modify it in some areas. The consequences might include severe drought or a significant change in mean temperature. This effect was considered by the Commission in their First Report([29]). The issue is still open, but if such climatic changes were to be firmly established the implications for energy strategy would be considerable([30]). Nuclear power offers an alternative which at least does not have this particular consequence.

193. The expansion of nuclear power and the consequent reduction in the use of coal would reduce the future need for mining, and hence for men to work in unpleasant, and sometimes dangerous, conditions. We must not forget that these problems also arise in the mining of uranium in other countries, although the use of fast reactors on a large scale would greatly reduce the mining operations required for a given energy supply.

194. Another advantage possessed by nuclear electricity is that it appears to be significantly cheaper, at least in the UK, than electricity from coal and oil power stations which has recently increased dramatically as a result of the much higher fuel costs. There is also advantage in having diversity of supplies, which makes the electricity grid less dependent on a single source of primary energy. Indeed the SSEB specifically mentioned this as an argument in favour of ordering more nuclear stations in the future. While we can see that to have a fair proportion of nuclear power stations would be advantageous, it would not necessarily be prudent to aim at a predominantly nuclear system for the very same reason.

195. The last point we would adduce is the fact that nuclear power stations use much smaller quantities of fuel than do fossil-fuelled ones. For example, one AGR station of 1,000 MW nominal output requires only 50 tonnes of fuel a year, whereas an oil-fired one would need the oil contained in twenty tankers each the size of the "Torrey Canyon". The transport of large quantities of fossil fuels presents environmental problems and, especially with oil, the risk of accidents that may cause severe environmental damage.

Conclusion

196. The advantages described above must certainly be weighed against the fears and the risks attaching to nuclear power, which lead many people to regard its widespread development as a grave and unacceptable menace to the world. Acceptance of this development and these risks in return for the promise of abundant electricity supplies has been called "the Faustian bargain". It needs to be considered how real these fears are. Certainly if we look at the experience and the present extent of nuclear power in this country it might well appear that in contemplating such fears we are creating unnecessary spectres. We are concerned, however, with the future and the fact that the world is on the threshold of a huge commitment to fission power which, once fully entered into, may be effectively impossible to reverse for a century or more.

197. We conclude that development of fission power on the scale we have described earlier carries implications and potential risks for society which are too serious to be disregarded on the grounds that they are necessarily speculative and of a kind that we have not hitherto expected to address in decisions on technological development. Decisions should not be taken simply on the basis of technological or economic advantage and the assumed necessity of securing steadily increasing energy supplies. The social and ethical issues involved are real and important, and should be widely appreciated and discussed. We reserve further consideration of these issues until Chapter X.

CHAPTER V

INTERNATIONAL AND NATIONAL CONTROL ARRANGEMENTS

Introduction

198. No-one who surveys the field of radiological protection can fail to be aware that the hazards of ionising radiation are well appreciated, and that there is, and has been for a long time, an elaborate system of national and international bodies which work together to minimise the potential dangers. There is a concomitant research programme and although, as we observed in Chapter II, there are still many detailed questions to answer, much has been learned. We have been impressed by the relatively much stricter regime that prevails with respect to ionising radiation than in the field of toxic chemicals. The result has been an exemplary record in the nuclear industry world-wide of protection of the health of both radiation workers and the general public. Indeed, we are strongly of the view that many of the practices for the protection of human health that are common in the radiation field could and should be adapted for application in other areas. To the extent that radiological protection plays a pioneering role, even greater resources can be justifiably allocated to it than would be appropriate to the needs of the immediate situation. We referred to one such practice, the study of a critical group of the individuals most exposed to a particular pollutant in order to determine safe discharge criteria, in our Fourth Report (paragraphs 144, 147).

199. Nevertheless, we are also aware that the organisational arrangements are complex, that in these circumstances the responsibility for different tasks may not be entirely clear, and that in the future, if there is a big expansion in nuclear power, qualitatively new problems are likely to arise and changes may be needed. It was these views that led initially to our decision to undertake the study. We have in this chapter set out the arrangements as they exist at present, we have made observations upon their efficacy, and we have made recommendations for such changes as appear to be desirable. We are not here concerned with whether there *should* be discharges of radioactivity to the environment, or of how they should be reduced, but rather with the arrangements for ensuring that the general public (and the natural environment) are protected and whether these are adequate.

Basic standards

200. The procedure of assuring protection for members of the public from a discharge of radioactivity has three stages. The first of these is to decide upon the maximum allowable dose of ionising radiation, either to the whole body or to some particular organ (see paragraph 56). The second step is to discover the connection between discharge of a particular radioisotope at a certain rate (e.g. 200 curies/month) and the radiation dose thereby received by the most

exposed member of the public or by an average member of a "critical group". This process has to take account of the possible environmental pathways, which can be expected to change with time (see paragraph 80). The third step is to decide on what discharge will actually be permitted, and on the monitoring and reporting system needed to ensure that the level is not exceeded. The permitted level must take account of all discharges so as to ensure that the total quantities of radioactivity in the environment are compatible with the maintenance of public health.

201. Basic radiation standards are recommended by a body called the International Commission on Radiological Protection (ICRP). This is a group of scientists, currently 12 in number, chosen every four years by the International Congress of Radiology on the basis of their individual scientific reputations and independence. There is no particular attempt to allocate places on a national basis, and in fact at present four out of the 12 members are British. The ICRP is independent of Governments, and is not within the framework of United Nations agencies; it is answerable to the world's professional radiologists meeting in congress, and its publications and recommendations are used by all major countries as a basis for protection against biological effects of ionising radiation.

202. The ICRP begin by regarding all increases of radiation as to some degree harmful, and therefore recommend that they should be kept as low as is readily achievable. They have also laid down, on the basis of estimated risks and actual experience, that doses to individual members of the public should not be greater than certain defined amounts. These doses are one tenth of those listed in Table 3, paragraph 56. The present standards have existed with only minor changes since 1958. They are kept under review, and a major exercise is now in hand with the aim of producing revised recommendations. The ICRP attempt to make the permitted maximum doses to the different parts of the body consistent, so that the level of risk associated with different types of exposure at the maximum allowable should be approximately the same. On the basis of the linear hypothesis, and the figures given in paragraph 52, a dose of 0·5 rem per year to the whole body might be expected to cause eventual death with a probability of about 1/20,000 per year.

203. The ICRP recommendations are expressions of opinion based on deductions from scientific fact. Clearly the ICRP is only as good as its members, and it is vitally important that these should continue to be appointed independently of the approval of their national governments and purely on the basis of their professional standing amongst their scientific peers. Given this, we can see no better way of deriving basic standards than by accepting the ICRP recommendations. We hope that there will continue to be a strong British representation on the ICRP as evidence that research work conducted here is of a high standard.

204. The ICRP also recommend secondary standards for particular radioisotopes. These can take the form of maximum permissible body burdens, maximum permissible annual intakes, or maximum allowable concentrations in air or

water. In general, each of the standards, like the basic table of allowable radiation doses, is ten times as strict for members of the general public as for radiation workers. These secondary standards depend on scientific studies whose interpretation is sometimes the subject of dispute; we have referred to one such matter in paragraphs 68–71. Such dispute is likely to be heightened if, as may happen, a change in a standard would require changes in operating practice that would have considerable economic penalties. There is then the need to balance economic factors against uncertainties in scientific evidence.

205. For the ICRP recommendations to be put into effect, they must first be endorsed as standards by national governments, and translated into codes of practice and detailed regulations to provide guidance for control authorities and operators. It is highly desirable that there should be international agreement and coordination on these practical matters, and it is therefore no surprise that a good number of international organisations of an official or governmental kind have grown up for this purpose. There are two main classes of organisation—United Nations and regional—and they have different functions and powers.

206. The United Nations Scientific Committee on the Effects of Atomic Radiation (UNSCEAR) is the prime scientific review body on a world basis. It was set up by the General Assembly in 1955, and consists of governmentally-approved representatives from 15 countries who meet annually in the headquarters of the IAEA (see below). It reports on the levels of radiation from different sources, and on the scientific evidence of their effects, and its publication of these in 1972 can be regarded as authoritative.

207. The International Atomic Energy Agency (IAEA) arose out of a speech on "Atoms for Peace" by President Eisenhower in 1953, and was founded as a UN Agency in 1957 to promote the peaceful use of atomic energy. It has over 100 member states who send representatives each year to the headquarters, in Vienna, where there is a permanent Secretariat. It has amongst its functions, the promotion of the peaceful uses of nuclear power and the establishment and administration of safeguards in connection with nuclear activities, including the transport of nuclear materials and the protection of fissile material. It acts as an intermediary for the supply of fissile material and it also holds a substantial number of technical and scientific symposia, and publishes their proceedings. There are IAEA basic radiation standards, which are based on ICRP recommendations. These are advisory for Member States, but binding on States (chiefly the less developed States) that receive IAEA materials, services or equipment under an agreement with the Agency. Such States are also subject to the detailed safeguards system of the IAEA which provides for inspection by IAEA Inspectors. Two other UN agencies with an interest in basic radiation standards are the World Health Organisation (WHO) and the Food and Agriculture Organisation (FAO). The WHO have published European (1970) and International (1971) standards for Drinking Water, and these include upper limits for radioactivity derived from ICRP publications.

208. There are two regional groupings of nations with an interest in nuclear power and radiation standards. These are the Nuclear Energy Agency (NEA) of the Organisation for Economic Co-Operation and Development (OECD), based in Paris, and the European Atomic Energy Community (EURATOM). The membership of OECD broadly comprises the Western European States together with the USA, Canada, Japan, Australia and New Zealand. The NEA has an active Secretariat and promotes a number of scientific conferences whose proceedings are published. It also runs a system of technical committees, including the Committee on Radiation Protection and Public Health (CRPPH) which meets twice a year and maintains a continuous review of radiation protection standards. NEA standards, like those of IAEA, are based closely on ICRP. Generally OECD/NEA is concerned to promote principles which are subsequently incorporated into the administrative and legal systems of member states. One activity of the NEA that is particularly relevant to our study is the organisation of radioactive waste disposal operations to the deep Atlantic on behalf of its Member States (including Britain). We discuss this aspect further in paragraph 369.

209. Unlike all the other international bodies described above, EURATOM does have powers to set and enforce basic radiation standards, by virtue of the Treaty of 1957 to which the UK acceded on joining the European Communities in 1973. Article 30 of the Euratom Treaty requires the Community to establish "basic standards for the protection of the health of workers and of the general public". These include maximum permissible doses, maximum permissible levels of exposure and contamination, and basic principles of the medical supervision of workers. These are also based on ICRP recommendations, but some of the secondary standards differ slightly.

210. The Treaty provides a mechanism for these standards to be devised and adopted. They are to be "worked out by the Commission after it has obtained the opinion of a group of persons appointed by the Scientific and Technical Committee from among the scientific experts, and in particular the public health experts, of Member States". The basic standards are to be referred to the Economic and Social Committee (on which industry and the trades unions are represented) and the European Parliament. Adoption of the Directives in which the standards are embodied depends finally on a majority vote in the Council of Ministers, and in contributing to the Council the British Government has to form a view on the soundness and acceptability of the proposals. We would emphasise the importance of ensuring that the Directives take full account of the ICRP recommendations.

211. The Group of Experts appointed by the Scientific and Technical Committee has 20 members of whom three are British. Since two of them are currently also full members of the ICRP and the third is alternate to one of the others on UNSCEAR, and since the same is likely to apply to some members of other nationalities, it may be seen that there is a fair amount of cross linking between these three bodies, and that it is to be expected that they will move in step.

212. The Directives are not themselves legal instruments and each Member State must thereafter take whatever legal or administrative steps are necessary to bring the standards into effect. Thus each nation in the Community has laws that limit the allowable doses of radiation to members of the public and provide means for control and enforcement.

Radiation standards in the UK

213. The UK enacted such legislation in the form of the Radioactive Substances Act 1960, which was based on a White Paper, "The Control of Radioactive Wastes" (Cmnd 884), 1959. Disposals of radioactive waste are permissible only when authorised under section 6 of the Act and are based on the following three criteria:

(*a*) to ensure, irrespective of cost, that no member of the public shall receive more than the relevant ICRP dose limits (e.g. 0·5 rem per year for the whole body);

(*b*) to ensure, irrespective of cost, that the whole population of the country shall not receive an average of more than 1 rem per person in 30 years (i.e. one-fifth of the ICRP limit relating to the genetic risk); and

(*c*) to do what is reasonably practicable, having regard to cost, convenience and the national importance of the subject, to reduce doses far below these levels.

The third criterion corresponds to the ICRP requirement (see paragraph 202) that all doses should be kept as low as is readily achievable. In fact, there is no difficulty in achieving an average population dose far below the level specified in (*b*). The criteria do not include radiation doses from other man-made sources such as consumer goods containing radioactive materials. Doses from all man-made sources (except those given for medical purposes) are however included in the ICRP basic standards.

214. The Radioactive Substances Act does not specify the basic radiation doses, but it does empower the Minister to make regulations; the 1961 Ionising Radiations (Sealed Sources) Regulations were so made. The regulations are based upon acceptance of the ICRP permitted doses for radiation workers and for the general public. Such regulations can also be made under the Health and Safety at Work &c Act 1974. These, being statutory instruments, have legislative force, and they will give effect to the Directives of the EURATOM Treaty.

Derived working limits and discharge limits

215. We have briefly described the international organisations that have been formed to deal with radioactivity and, in particular, the central role of the ICRP in formulating the basic standards which define dose limits. Before considering the organisational arrangements in the UK it is convenient to introduce some other underlying features of control. We noted in paragraph 200 that basic standards must be translated into practical limits for application to

particular radioactive discharges. Control implies the establishment of a quantitative relationship between the rate of discharge from a specified source and the resulting radiation dose to man. This depends on knowledge of the pathways followed in the environment by the radioisotopes concerned, so enabling identification of the critical pathway, corresponding to the most significant route back to man, and of the critical group of the population who would receive the highest doses. Given the relationship between discharge and dose, Derived Working Limits (DWLs) can be formed from the basic standards; these are defined by the ICRP as limits derived from the basic standards such that compliance with the DWLs implies virtual certainty of compliance with the relevant dose limits. The closeness with which the link between DWL and the dose limit can be established depends on circumstances. For example, if a radioactive material is incorporated to a known degree in a foodstuff, and if the eating habits of the population involved with respect to this foodstuff are known, then the link can be closely determined. On the other hand, if we were concerned with the inhalation of radioactive material discharged to air, the derivation of the DWL would be complicated by uncertainty about the degree of dispersion and a margin of safety would need to be included to cover this.

216. The determination of DWLs depends on scientific knowledge of the pathways, of the distribution of radioactive materials in the environment and of their possible accumulation in food chains. The pathways of importance will change with time; for example, as a result of changes in dietary habits or changes in scientific understanding of the mechanisms involved. For radioactivity, as for other forms of pollution, effective control arrangements must rest on continuing scientific investigation and research. The arrangements for initiating and co-ordinating research and for ensuring interaction between this activity and the control processes are matters of great importance.

217. We have noted that although firm dose limits are specified, there is a general requirement to ensure that all doses are kept as far below these limits as can reasonably be achieved. Therefore the actual discharge limits set for particular discharges will reflect other than purely scientific considerations and must be arrived at by judging the benefits of reduction of exposure below the dose limits against the costs involved in securing the reduction. Thus, economic, social and technical factors are involved, and this has implications for the control machinery.

Emergency reference levels

218. The limits we have described above apply only to routine discharges made under normal operational conditions. There is the possibility that the public in the vicinity of a nuclear plant could be subjected to greater exposures arising from accident conditions. This could be such as to call for intervention by the authorities, for example, to limit access to particularly affected areas, to distribute medication or to evacuate some areas. Such action would cause hardship and economic penalties as well as some risk, and the need for it must therefore be weighed against the risk arising from the radiation itself. The desirability of providing guidance to the authorities in dealing with such

situations has led to the specification of Emergency Reference Levels (ERLs). In effect, these levels specify minimum radiation doses which, if likely to be sustained as a result of an incident, would call for counter-measures to be considered. The nature of these measures, and whether they should be implemented, would depend on the extent to which the ERLs might be exceeded and on an assessment of the risks of the counter-measures to the community concerned.

Arrangements for radiological protection in the UK

219. Our concern is with the control of radiation hazards arising from nuclear power. As we have noted, the organisational arrangements are somewhat complicated and we thought that it would assist understanding of the subsequent discussion to introduce the various agencies involved, and to illustrate their functions, in diagrammatic form. This is done in Figure 12. The Nuclear Installations Inspectorate (NII) and the Factories Inspectorate (FI) have been included in the figure for the sake of completeness, but we reserve consideration of the NII's role until the next chapter. The FI's concern with radioactivity is limited to industrial applications and we do not consider this aspect further. We should comment on the position of HM Alkali and Clean Air Inspectorate (ACAI). This Inspectorate is at present part of the HSE, though answerable through the Health and Safety Commission to the Secretary of State for the Environment on questions of environmental policy. In our Fifth Report([31]) we recommended that the ACAI should return to direct control by the DOE and that its functions should be extended so that it would be in a position to deal comprehensively with all forms of pollution arising from difficult industrial processes. We called this new body ("Her Majesty's Pollution Inspectorate" or HMPI and we shall have occasion to refer to it again later in this chapter.

The endorsement of basic standards

220. We have described the roles of the ICRP and EURATOM in devising basic standards which provide the foundation for control measures. The ICRP recommendations are universally regarded as authoritative but in a matter of such fundamental importance there is a need for independent assessment, not least as a safeguard should ICRP members ever come to be determined more by government choice than by professional achievement. It is a clear requirement that there should be an expert body having a statutory responsibility to advise the Government on whether the basic standards that are proposed are acceptable. We were somewhat surprised to learn that there is, strictly, no such body at present. The Medical Research Council (MRC) is in fact held responsible for advising the Government on the biological implications of these standards and on their acceptability in the UK; this responsibility was expressed in a letter sent from the (then) Secretary of State for Health and Social Security to the Chairman of the National Radiological Protection Board (NRPB) at the time when the Board was formed (see below). However, we were advised in MRC evidence that the Council had no formally defined or statutory responsibilities in the field of radiological protection. Such a loose arrangement with regard to so basic a matter would scarcely be satisfactory for the future.

FIGURE 12

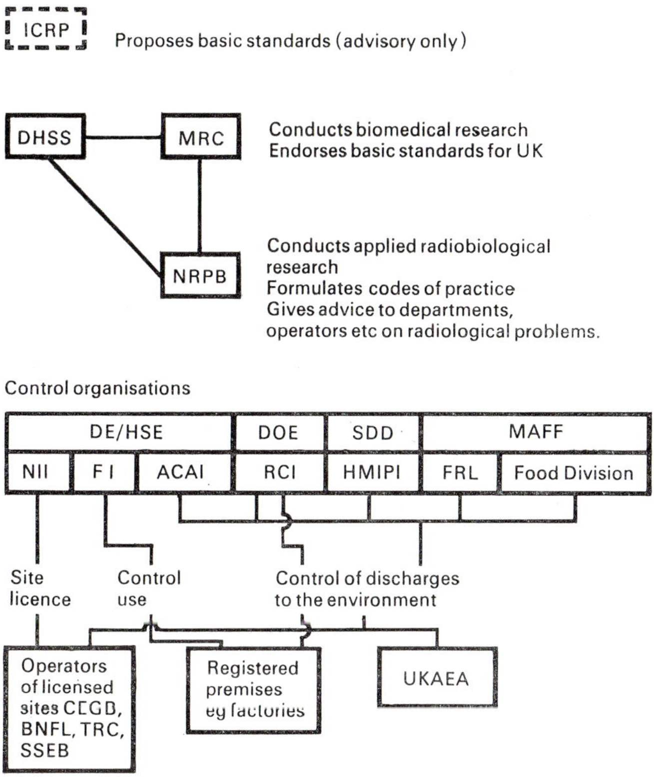

Key:	
ICRP	International Commission on Radiological Protection
DHSS	Department of Health and Social Security
MRC	Medical Research Council
NRPB	National Radiological Protection Board
DE	Department of Employment
HSE	Health and Safety Executive
DOE	Department of the Environment
SDD	Scottish Development Department
MAFF	Ministry of Agriculture Fisheries and Food
NII	Nuclear Installations Inspectorate
FI	Factories Inspectorate
ACAI	HM Alkali and Clean Air Inspectorate
RCI	Radiochemical Inspectorate
HMIPI	HM Industrial Pollution Inspectorate
FRL	Fisheries Radiobiological Laboratory

Diagram showing the organisation of radiological protection in the UK.

221. We referred above to the NRPB. This body was set up in October 1970 in accordance with the provisions of the Radiological Protection Act 1970. It subsumed the functions of the Radioactive Substances Advisory Committee (set up under the Radioactive Substances Act 1948) and took over the Radiological Protection Service which had been run by the MRC. The Board's statutory functions are to carry out research into radiological protection and to provide information and advice on the subject to those having responsibilities in the field, including government departments. It was intended that the NRPB would provide "a national point of authoritative reference in radiological protection".*

222. To judge from its title, one could expect that the NRPB would occupy a focal position in the organisation for radiological protection. In fact, from our discussions with the departments and agencies specifically concerned with nuclear power we gained the impression that they regarded the Board as having, in this field, a somewhat peripheral role. No doubt this is in part due to the fact that the Board is still a relatively new organisation whose relations with other bodies are still developing. Another factor, however, is probably that these bodies generally possess considerable knowledge in radiological matters and are unlikely to feel that they need often to turn to a body whose function is essentially advisory. The practice of consultation will undoubtedly develop as the Board's relations with other bodies become established but we think it is right that consultation should become a formal requirement, the more so in view of the wider responsibilities that we believe the Board should assume and which we discuss below. We therefore recommend that the Government should make consultation mandatory at a suitable and early date.

223. Our present concern is with the question of which body should bear the statutory responsibility for endorsing basic standards for radiological protection. Such a body must clearly be independent of any interests which are concerned with promoting developments involving radioactivity, and in view of the importance of the issues it should be a body which has been constituted specifically for radiological protection. We are fully agreed that the NRPB should be developed to form this body. The NRPB is administratively answerable to the Secretary of State for Health and Social Security jointly with the Secretaries of State for Scotland and for Wales on appropriate matters. There is at present a requirement embodied in the Radiological Protection Act 1970 that these Ministers should consult with the AEA and the MRC with regard to any directions they may give to the Board, and on appointments to the Board. We can see that consultation with the AEA on such matters may be useful but we think it is inappropriate that it should be mandatory, given the future role that we envisage for the Board. We recommend that appointments to the Board should be made by the Ministers concerned in consultation with the MRC and the Royal Society.

224. We wish to make perfectly plain our view that for the NRPB to assume satisfactorily the function described above, as well as the responsibilities referred to in paragraph 237, will imply change within that body. For example,

*Stated by the then Minister of Health when introducing the Bill to Parliament.

in paragraphs 74–5 we commented on the NRPB's role in establishing the facts relating to the incidence of leukaemia in plutonium workers at Windscale. We do not think that the NRPB followed up this matter initially with the necessary thoroughness. We are clear that the changes we recommend in NRPB responsibilities are sufficiently far-reaching as to call for its reconstitution at Board level and for a review of the organisation and expertise of the executive body. We would again stress the importance we attach to NRPB independence; specifically, its independence from the AEA. Many of the present staff were previously with the Authority and indeed the NRPB took over some activities that were formerly carried out by the AEA; the NRPB is based at Harwell. Consequently, the NRPB is quite widely identified in the public mind with the nuclear industry. In making these remarks we intend in no way to impugn the integrity of the present members or staff of the Board, but rather to emphasise the importance of fostering not only the reality, but the appearance, of independence.

225. The changes we propose in the role of the NRPB will carry implications for the Department of Health and Social Security (DHSS). We have not considered this matter in detail and we did not receive any evidence from the DHSS. We foresee, however, that the Department may need to acquire additional expertise if it is to discharge its responsibilities adequately. We recommend that this question should be considered by the Government.

226. We envisage that the close contacts that already exist between the NRPB and the MRC and, in particular, their Joint Committee on Radiological Protection, will continue. The MRC would be free to criticise NRPB or DHSS on radiation standards but would have (as now) no statutory responsibility. The Council would of course continue at its discretion to conduct research into the scientific basis of radiation standards.

Research

227. We have commented on the importance of research which provides the scientific foundation for the control of radioactivity. Several different areas of research can usefully be distinguished. The first area is biomedical research which is concerned with determining the effects of exposure and the treatment of people who have been dangerously exposed. Another area is concerned with the general environment and the determination of how radioactivity is dispersed and concentrated until it may have an effect on man and other forms of life. A third area is concerned with techniques for measuring and monitoring the levels and effects of radioactivity and with assessments of the radiological properties of particular materials. We have considered whether the arrangements to promote research in these different areas are adequate, whether there is the right degree of support and whether there is sufficient feedback between research and practical application.

228. Basic radiobiological research is primarily funded through the MRC which not only supports independent work in universities but also maintains its own Radiobiological Units at Harwell and elsewhere. Despite the proximity of the Harwell Unit to the NRPB and to the AEA's Health and Environmental

Group at the Atomic Energy Research Establishment (AERE), we have formed the impression that there has been insufficient coordination of radiobiological work. We therefore welcome the recent setting up by the MRC and the NRPB of a Joint Committee on Radiological Protection to improve liaison. One of the Committee's tasks should be to consider any recommendations for research work that arise from the discussions within ICRP and to ensure that, where it seems appropriate, the work is done in the UK, and that there is agreement on which institutions should be responsible for pursuing it.

229. The research conducted by the NRPB is largely "applied" in character and directed to the investigation of effects and techniques that are relevant to the diagnosis and treatment of exposure in radiation workers. Such research should remain the responsibility of the NRPB, who have built up a good reputation in a short time for some of their work in this field. In view of the possibility of the large-scale use of plutonium in a future nuclear power programme, the NRPB is giving some priority to research on plutonium and the higher actinides. The Board also supports some research in other establishments and universities.

230. The view was expressed to us by both the MRC and the NRPB that the resources currently being devoted to radiological research in their respective fields were about right and that the most important tasks were being adequately covered. Generally, we accept that this is so.

231. Our next concern is with research into the effects of radioactivity in the environment and into its pathways back to man. We need to consider the atmospheric, terrestial and aquatic environments and we take the latter first since the arrangements for research in this field are most clearly defined. For normal operational discharges of radioactivity the main potential impact of releases to atmosphere is on agriculture and that of releases to sea is on fisheries. The Department principally concerned is therefore the Ministry of Agriculture, Fisheries and Food (MAFF).

232. As there is no counterpart on the fisheries side of the Ministry's responsibilities to the Agricultural Research Council (ARC), MAFF have undertaken radiological research in the aquatic environment (primarily marine) directly through their Fisheries Radiobiological Laboratory (FRL) at Lowestoft. We have visited this laboratory and we were struck by the range of work undertaken and by the thoroughness with which the laboratory discharges its responsibilities. The staff not only conduct research on how radioactivity is distributed in the environment and returns to give radiation doses to certain critical groups, but also inspect the sites and determine the levels of discharges that should be authorised. It appeared to us that this dual role of research and inspection had been fruitful in helping to identify useful areas for future research and in bringing expert scientific knowledge to bear directly on control problems.

233. Though we were impressed by the quality of the research conducted by the FRL we think it is important that the Government should encourage

independent work in this field elsewhere, through the Research Councils*. We are also concerned that sufficient resources should be made available for adequate investigation of the new problems that could arise in the future, given a large expansion in nuclear power.

234. We are much less content with the position on research on radioactivity in the atmospheric and terrestial environments. Much of the work on radioisotopes in agriculture and food was carried out in the 1950s and 1960s at the ARC Letcombe Laboratory and was stimulated by public concern about fallout from atmospheric tests of nuclear weapons. Since the Partial Test Ban Treaty of 1963 the levels of radioactivity both in the environment generally and in milk (which is the most important source of human exposure) have declined to quite low levels. The emphasis on radioactivity in the laboratory's research programme has accordingly been reduced. Within MAFF, responsibility for radioactivity in relation to food and agriculture is exercised by the Food Science Division, which issues (jointly with DOE) authorisations for routine discharges to atmosphere from nuclear sites. In contrast to the position on the fisheries side, however, the Division has no research capability.

235. We have reached the conclusion that insufficient research is currently being done on radioactivity in the atmospheric and terrestial environments. There is especially a need to investigate in good time the problems that could arise with expansion in nuclear power and with the introduction of new types of plant which could lead to releases, whether planned or accidental, of new kinds of radioisotopes. For example, it is important that there should be adequate work on atmospheric diffusion, deposition and re-suspension of aerosols, as these may provide some of the most significant routes for radioisotopes in the future.

236. The question arises of what arrangements could best be made to this end. One possibility is to extend the remit of the FRL. As we have seen, this laboratory has demonstrated considerable competence in its present field and it already does some work in atmospheric monitoring. However, we have judged that this would not be the right approach. The work required on the terrestrial environment covers too wide a range of scientific disciplines and needs to be done by expert ecological and medical groups in appropriate research establishments and universities. We think that the nature of the research is such that it would fall mainly within the remit of the Natural Environment Research Council (NERC) and that this Council should be funded to undertake this work. There would need to be consultation with the ARC and the MRC and with the NRPB (see below). There should of course be contact between this research programme and that of the FRL.

237. Additional research, such as might not qualify for support by the Research Councils, should be commissioned and financed by the NRPB. The Board should have responsibility for the broad programme of environmental research on radioactivity, and for ensuring that necessary research is undertaken having regard to expected developments in nuclear power and the nature

*One area in which we regard such work as particularly important is that of sedimentology, in relation to discharges of plutonium from Windscale (see paragraphs 352–354).

and quantities of wastes that may need to be discharged in the future. The NRPB should collaborate with the NERC and other bodies referred to above in identifying the research required. It is an important requirement that research contracts should be open, with freedom to publish subject only to patent and security considerations. The Board should also take account of relevant research work undertaken by the AEA, BNFL and the utilities; for example, work done by CEGB on the diffusion of radioactivity in the atmosphere.(7)

238. We have so far been concerned with research aimed at establishing the effects of radioactivity on man and its environmental pathways back to man. We noted in paragraphs 64 and 81 (Chapter II) that there appear to be grounds for supposing that radioactivity poses a lesser threat to other forms of life. It is pertinent to recall, however, that early work on pesticides was based on direct mammalian and human toxicity figures which suggested that chlorinated hydrocarbons were harmless, but that it was subsequently found that these could accumulate in and harm other species near the top of the food chain. There is a need for continuing research to assess the effects of radiation on the natural world.

The control of discharges

239. We turn now to the arrangements for controlling discharges of radioactivity. The legislative basis for control is provided by the Radioactive Substances Act 1960. The responsibility for authorising discharges is held jointly by the Secretary of State for the Environment and the Minister of Agriculture, Fisheries and Food for licensed nuclear sites and those of the AEA. In Wales, the responsibility is held jointly by the Secretary of State for Wales and the Minister of Agriculture, Fisheries and Food though the DOE Radiochemical Inspectorate (RCI) act for the Welsh Office on technical matters. A corresponding arrangement for joint departmental responsibility would apply in Northern Ireland but there are at present no nuclear sites in the province. In Scotland, the Secretary of State for Scotland is solely responsible. For simplicity, and since we are concerned mainly with the principles involved, we refer mainly to the arrangements in England in the following discussion.

240. The arrangement (in England and Wales) for joint responsibility for authorisations is at first sight curious. The involvement of MAFF reflects particularly their interest in fisheries and the fact that the major discharges of radioactivity are to the marine environment. The arrangement calls for close liason between the units involved in DOE and MAFF and seems administratively awkward, but it appears to have worked satisfactorily in practice. Discharges to atmosphere are authorised by HM Alkali and Clean Air Inspectorate (ACAI) for the DOE and by the MAFF Food Science Division. For discharges to water (mainly to coastal waters) there is an agreed division of effort between the two Departments, the RCI dealing primarily with matters relating to effluent control while MAFF, through the FRL, are mainly concerned with assessing the environmental consequences of the discharges. Discharges to land, that is, burials of radioactive material, are sanctioned by the DOE on the basis of established rules to govern disposals to ordinary refuse tips, and tips

where "special precautions" are applied. Disposals from licensed nuclear sites onto two special tips (Drigg, Cumbria and Ulnes Walton, Lancashire) operated by BNFL which can accept higher levels of activity than the others, require a joint authorisation from DOE and MAFF. In Scotland, HM Industrial Pollution Inspectorate (IPI) issues authorisations for discharges to air, water or land on behalf of the Scottish Development Department (SDD) after consultation with the Department of Agriculture and Fisheries for Scotland (DAFS) about their interests. (DAFS, however, authorise the sending of solid waste for ocean dumping). With regard to discharges to the sea, the IPI work closely with the FRL to whom DAFS have delegated the work of carrying out coastal radioactivity surveys round Scotland.

241. We have described how Derived Working Levels (DWLs) are determined from the basic standards, so providing the basis for establishing discharge limits, and the dependence of this process on research on environmental pathways. In the UK at present this process is carried out by the Government Departments who issue the authorisations, in contrast to the procedure in most other countries within the European Community where the responsibility falls on the operators of nuclear plant. Their proposals for discharges and supporting analyses are inspected by government departments and, in some countries, made subject to public hearings. It appears right in principle that operators should bear the onus of responsibility for justifying their releases of radioactivity and showing that these will be innocuous or acceptable. This, however, might lead to duplication of effort, for the final decisions must be taken by the controlling authorities and if this is to be done responsibly they will require scientific knowledge and expertise which can be ensured only by undertaking related research. We conclude, therefore, that the primary responsibility for assessing the effects of discharges on the environment and man should continue to rest with Government Departments. This responsibility should be independent of those of Government Departments formulating energy policy or licensing nuclear sites.

242. We have already expressed reservations about supporting research in some areas. Subject to this we are broadly satisfied that the present arrangements, whereby the DOE and MAFF (with their respective responsibilities for the quality of our air and drinking water and for our food) have joint responsibility for deciding the DWLs for discharges of radioactivity, are adequate for the present level of nuclear development. Discharges to the environment are generally quite low and any individual member of the public will usually receive a significant radiation dose by only one route. If there is a large expansion of nuclear power in the future, however, the position will become more complex and some people may receive radiation doses from more than one source sufficient to require their additive effects to be considered. There will then need to be a mechanism to ensure that the controlling authorities take account of such effects. It must be remembered, too, that the half-lives of some radioisotopes are long and that each release implies a future dose commitment. The long term implications of cumulative effects of radioactivity releases must be estimated and kept in mind.

243. It is in relation to these possible future developments that we feel some anxiety about the present arrangements. The arrangements appear somewhat fragmented and they do not give us sufficient confidence that the effects of an expanding nuclear programme will be comprehensively investigated and assessed; they lack a focus. We recommend that the NRPB should be given this central role. This is indeed implied by the recommendation we have made (paragraph 237) that the Board should be responsible for keeping under review the total programme of environmental research on which the DWLs and control are ultimately based, and for ensuring its adequacy in relation to likely developments in nuclear power.

244. We have explained that the limits set for particular discharges reflect the general requirement that doses should be kept as far below the ICRP dose limits as can reasonably be achieved. Discharge limits are commonly expressed as a proportion of the relevant DWLs. In deciding what this proportion should be for any particular discharge, the risk to the community who may be affected must be set against the implications for the operator of the limit that is specified. Thus, a requirement that a discharge should be reduced from, say, 400 curies of β–activity per month to 200 might imply substantial costs for additional equipment. It would also mean that more radioactivity was retained on the site, perhaps adding to the hazard to operating staff. These effects would need to be weighed against the reduced risk to the public.

245. Under present arrangements discharge limits are set as an extension of the process of determining DWLs. Thus, the ACAI, in consultation with MAFF Food Division, control emissions to atmosphere. The limits for discharges to water are determined largely by the FRL who negotiate with the plant operator on the discharges that should be allowed, taking costs and other factors into account. We think that these arrangements for deciding the limits that apply to discharges of radioactivity in different forms may not lead to the optimum solution. The reduction of a particular discharge does not eliminate the radioactivity but merely changes it into another form. Thus, if extra filters are used to reduce an emission to atmosphere there will result a correspondingly larger mass of moderately active material to be stored or buried. Discharges to water can be treated with ion-exchange resins, but these then pose troublesome disposal problems. Burial of radioactivity on a landfill site may carry the risk of contamination of groundwater, or of resuspension of aerosols in dry and windy weather. Alternatively, if the site is revegetated, radioisotopes may be incorporated into the crops.

246. This notion of the transferability of pollution is one that we discussed fully in our Fifth Report(31) in the context of industrial pollution generally. We there argued that where industrial processes gave rise to difficult pollution problems, the responsibility for control of pollution in all its forms should be vested in one body (HM Pollution Inspectorate or HMPI) which would then be in a position, having taken account of economic, social and other relevant factors, to determine the "best practicable environmental option" for each plant. We recommended that this new Inspectorate should form part of the DOE and that it should be founded on the present ACAI.

247. We can see no reason why the same system of control should not apply to nuclear plant. For this purpose HMPI should incorporate the staff of the RCI and it should take over the inspecting, authorising and negotiating functions at present exercised by the MAFF Food Science Division and the FRL. In determining discharge limits, however, HMPI should be required to consult MAFF and the NRPB. We commented in paragraph 232 on the value of the present system in which certain scientists have a dual role of research and inspection. We would emphasise that in proposing the above changes we intend in no way to affect the rights or opportunities of research scientists to visit nuclear plants, or their access to relevant data. On the contrary, we would wish to see the closest possible contacts between the scientists and the HMPI staff involved. HMPI should consult the scientists on the monitoring that should be undertaken by operators (this is, and should continue to be, part of the conditions attached to the authorisation) and this should take account not only of the immediate need to protect public health but of research requirements. We note finally that the arrangements we propose above closely parallel those that already exist in Scotland.

Monitoring and surveys

248. Although radioactivity cannot be detected with any of the senses, it is much easier, generally speaking, to monitor in the environment than are many other pollutants. The instrumentation needed is relatively simple, unless a breakdown into the quantities of each radioisotope present is needed, whereas the accurate determination of heavy metals or complex organic compounds is usually fairly challenging.

249. There is a statutory obligation on operators of licensed sites to monitor various aspects of the neighbouring environment. The results are made available to local liaison committees but are not usually published as such. In fact, around virtually all the CEGB nuclear power stations the results show levels of radioactivity that are indistinguishable from background levels. The main interest attaching to the results in the past was the detection of the radioactivity attributable to fallout from nuclear bomb tests. With the passage of time since the Test Ban Treaty in 1963 the level has consistently declined. However the monitoring arrangements must of course be maintained, not least to ensure the detection of any accidental release that might occur.

250. The situation is rather different for the BNFL works at Windscale, from which the discharges of radioactivity (at least those to water) are bigger than those from all the other sites added together. This is not to say that they are excessive, but it does mean that great care must be taken that all the possible environmental pathways that could be significant in terms of human radiation exposure are evaluated. To guard against the possibility of routes being overlooked—either from Windscale or anywhere else—it is necessary to conduct general monitoring of human food, water and air, and the built environment. It is also desirable to check the levels of radioactivity in people, by whole-body monitoring and by analysis of excreta.

251. At present regular surveys are prepared and published by the FRL covering discharges to the aquatic environment. Bearing in mind the possible

expansion in nuclear power, we wish to see this practice extended. There should be an annual survey of discharges of radioactivity whether to air, water or land and in the new arrangements we have recommended this should be prepared by HMPI in consultation with other bodies involved, and in particular with the FRL. The survey should describe the factors that have been taken into account in determining the discharge limits that should apply. It should also include tables of results from previous years so that general trends may be more readily appreciated; the lack of such comparison tables from the present FRL reports makes extrapolation more difficult.

252. A more comprehensive though less frequent report should also be prepared by the NRPB. The report should include information on the total radiation background, including that occurring naturally and the amounts being added for medical purposes. It should review the position on radiation doses attributable to nuclear sites, including an account of doses received in this country as a result of nuclear activities abroad, and conversely. The report should also review current thinking on environmental pathways and discuss the dose commitment arising from past and projected discharges. The NRPB should be responsible for initiating and co-ordinating the general and periodic monitoring that is required in order to check that environmental pathways are not overlooked, and information on such activities should be included in its report. We envisage here the sample monitoring of people by Area Health Authorities, of water by the relevant water authorities, and of air by local authorities (which might, however, delegate this function to outside bodies possessing the necessary equipment and expertise).

253. In preparing its report the NRPB will need to consult the nuclear industry and employers of radiologists and radiographers to obtain records of doses received by radiation workers. We have already commented (paragraph 76) on the need for a complete national register of these workers, with a section comprising those exposed to plutonium and the higher actinides and including workers who have retired or left the industry. This is a task which we expect the NRPB to undertake. We note here that data are needed not only on workers exposed under routine operating conditions but also on those receiving accidental exposures, particularly by the incorporation of radioisotopes. The present requirement to report any such significant accident to the NII should be coupled with a requirement to refer each case to a co-ordinating panel of experts for advice on the best treatment options as well as on the follow-up investigations that should be undertaken. Such a co-ordinating body would thus be in a position to monitor the frequency and severity of accidents and ensure the best use of the human data available to increase knowledge of the radiotoxicity of incorporated isotopes. This group could be set up as an *ad hoc* one at present, but with the build up of medical expertise within the NRPB it could eventually be incorporated into that body.

Accidental releases of radioactivity

254. We have so far considered the arrangements for controlling routine discharges of radioactivity. There is the possibility that more substantial releases could occur under accident conditions and arrangements must be made to

deal with the situations that could then arise. We referred to one aspect of this, namely the setting of ERLs, in paragraph 218 and we recommend that the responsibility for specifying these levels should be clearly vested in the NRPB.

255. When a site for a nuclear power station is selected, there is subsequently planning control over developments within ⅔ mile that might react on the station, and on population increases up to 2 miles away. The intention is to prevent a site, once selected according to a numerical population criterion, subsequently becoming unacceptable. It is equally to ensure that suitable arrangements can be made to warn the local population in the event of a significant release of radioactivity, to distribute stable iodate or iodide tablets (see paragraph 56), and, if the magnitude of the release should justify it, to evacuate the population. At each station, an Emergency Reference Plan is drawn up which has to provide for complete evacuation of the population within ⅔ mile in the space of 2 hours, and for suitable monitoring arrangements for foodstuffs over greater distances. These plans are discussed with the Station Liaison Committees that are formed at each licensed nuclear site, and rehearsed from time to time with the emergency services. We recommend that such plans should be made available for inspection by interested members of the public who might be affected by them. We have been struck by the efforts that have been made to involve the local populations in an understanding of the issues involved, and consider that the arrangements are likely to be as satisfactory as could reasonably be devised. They are in marked contrast with the general lack of public consultation at potentially dangerous non-nuclear plant, which often causes substantial local anxiety.

256. A release of radioactivity from a power station site would be the responsibility of the operator, who would co-ordinate the emergency services. Emergency plans are also required for the transport of particularly hazardous radioactive materials which arise mainly from the nuclear power industry. There are, however, many routine shipments of radioactive material each year, mostly unconnected with the nuclear power industry, for which special emergency arrangements would not be justified. An accident during such a shipment, if it involved the civil authorities, would be classed as an incident under the National Arrangements for Incidents involving Radioactvity (NAIR), which is administered by the NRPB. Under this scheme, which was set up in 1964, there is a national list of bodies who have the expertise and resources to cope with a real or suspected leak of radioactivity in each area of the country. There is a surprising number of small incidents, for example 22 in 1975, and although most of them turn out to be trivial* the arrangements are carefully worked out and reports are made.

257. The various authorities concerned with the control of routine discharges are all involved with other relevant authorities in assessing the possible consequences of nuclear accidents and in devising appropriate emergency plans. Clearly these must reflect a compromise, taking account of the likelihood and severity of possible accidents and the problem of ensuring the effectiveness of the arrangements should an emergency arise. Our general conclusion is that the

*Out of 157 incidents to date, 54 have not actually involved any radioactivity.

administrative and practical arrangements that have been made for dealing with accidental releases, both large and small, are good and we would wish to see a like amount of careful attention being paid to other hazards, particularly from toxic chemicals.

Conclusion

258. In this chapter we have been concerned with the organisational arrangements for the control of discharges of radioactive wastes arising from nuclear power. We would emphasise that we have not here considered the arrangements that exist to deal with strategic questions about radioactive wastes; for example, to formulate policy regarding the future disposal of the high activity waste that results from fuel reprocessing. In this area we think that there are serious deficiencies in the present arrangements, but we consider these in the context of a full discussion of the problems of waste management in Chapter VIII.

259. Given the present extent of nuclear power we have found nothing seriously wrong with the present control arrangements which, in spite of their apparent complexity, have on the whole worked very effectively. However there are some areas of weakness and the changes we have recommended will, we believe, lead to the development of a more coherent organisation which will give greater confidence that the problems that may arise from substantial nuclear development in the future will be adequately investigated and countered.

CHAPTER VI

REACTOR SAFETY AND SITING

Introduction

260. We referred briefly to the subject of reactor safety in Chapter IV where we contrasted the risks of reactor accidents with those arising from other activities and events. The advancing scale and complexity of technological developments of all kinds tends to increase both the extent of the possible consequences of serious accidents and the problems of foreseeing and preventing the mechanisms by which they might be caused. Clearly, absolute safety cannot be assured. All that can be expected is that the techniques and disciplines used to ensure safety in the design and operation of a plant are such that the risk of accident is reduced to acceptably low levels. By this we would mean that the risk should be acceptable in relation to the benefit conferred by the plant.

261. Our concern is with the environmental effects of possible reactor accidents. We are certainly not in a position to assess the technical details of safety measures that are adopted in reactor design and operation, but we thought it necessary, in view of our environmental remit, that we should enquire into the principles that are applied in seeking to ensure reactor safety, and into the related organisational arrangements. We are particularly indebted to Mr. F. R. Farmer, former head of the AEA Safety and Reliability Directorate (SRD), who acted in a personal capacity as our consultant on these issues.

The effects of radioactivity releases

262. We have seen that nuclear reactors contain a vast amount of radioactivity, mostly associated with the fission products, and that elaborate precautions are taken in reactor design to prevent any accidental release. It is, however, generally possible to postulate failure conditions under which such releases could occur. Most concern centres on two issues—whether in the event of a potentially dangerous situation developing the reactor can be shut down quickly, and whether the residual decay heat from the fission products can then be removed. If the safeguards provided to cover these contingencies were to fail, the temperature might rise to values at which the fuel would melt and interactions involving the molten fuel within the core could develop pressures that would be sufficient to rupture the containment, so allowing a release of the gaseous and more volatile fission products. Some part of the volatile fission products would condense on the reactor structure but some would escape to the atmosphere.

263. Because of the violent nature of the processes involved it is extremely difficult to predict the magnitude of the releases that could occur. The view has

been expressed to us that a large release would be likely to exceed a few per cent of the gaseous and volatile fission products though most unlikely to reach 100 per cent, and that it would probably lie in the range 3 to 30 per cent. It is likely that beyond a certain point the consequences of a release would not be proportional to its size, since there are strong indications that for large enough releases the self-heating of the fission product cloud would cause a thermal uplift and increase dispersal, so reducing ground level doses and deposition. In the discussion that follows we shall assume a 10 per cent release of the gaseous and volatile fission products from a large (1000 MW) reactor on a typical semi-urban UK site. The self-heating effect described above could well occur for such a release, but it has not been allowed for in the account we give of the possible consequences, which could therefore be somewhat less serious than we describe.

264. The hazard to health posed by an accidental release can be fairly represented by the amount of iodine–131 involved (half-life, 8 days). The inhalation of radio-active iodine, or its ingestion in milk from cows that have been grazing on land on which it has been deposited, would give a radiation dose to the thyroid gland. (The inhalation of 3μCi (3×10^{-6}Ci) or the ingestion of 2μCi would give an adult the ICRP dose limit of 3 rem). Another important element in the release would be caesium–137. This has a half-life of 30 years and would cause prolonged contamination of land or buildings on which it was deposited. A large reactor would contain within its core at equilibrium about 10^8Ci of iodine–131 and about 10^7Ci of caesium–137. For an assumed 10 per cent release, therefore, there would be respectively 10^7Ci and 10^6Ci of these isotopes.

265. The actual effects of a release of radioactivity would depend greatly on the wind direction at the time. We may illustrate this point by reference to data for one existing UK reactor site giving the population in 30° sectors around the site out to a distance of 24 km. The site is on the coast and four sectors out of the 12 cover the sea. In five sectors the population is about 20,000, in two 40,000 and in the remaining sector 240,000. The influence of wind direction on the numbers of people at risk is clear. For this reason it is not easy to give a simple presentation of estimates of the consequences of a given release from a typical site. For most reactor sites in the UK there is a 50 per cent chance that if a release were to occur the wind would be blowing towards the sea, when the effects would be minimal. Thus, estimates must be expressed in probability terms.

266. Direct radiation from the radioactive cloud passing downwind would cause acute medical effects (with some deaths) among people fairly near to the site, including operating personnel, and might cause some eventual deaths from leukaemia. The inhalation of iodine could cause damage to the thyroid, leading to thyroid cancers, in people at distances up to 16–24 km from the reactor.

It has been estimated that there is a 30 per cent probability that the number of such cases would lie between 100 and 1000, and a 20 per cent probability that it would lie between 1000 and 10,000. Thyroid cancers would develop over a period of 10–20 years after the event and a small fraction of them would prove fatal. Other eventual deaths might occur from lung cancer due to the inhalation of another element, ruthenium, which is insoluble and would give a radiation dose to the lung. The most probable outcome would be 100–150 deaths from thyroid cancer and 10–200 deaths from leukaemia and lung cancer, but these figures could be an order of magnitude greater or less depending on the conditions.

267. The passing radioactive cloud would deposit fission products on the ground and other rough surfaces, which would contaminate milk and foodstuffs and cause direct radiation to people in the district. The first of these effects can readily be controlled but the second presents more difficulties. It is expected that the radiation would arise mainly from deposited iodine, caesium and some ruthenium. The iodine would decay within a few weeks but caesum is known to attach itself to some soils found in the UK in such a way that 30–40 per cent would be retained in the top few centimetres of the ground. The radiation from this caesium might continue with an effective half-life of 20 years though it could probably be substantially reduced by deep ploughing. It is not clear how quickly deposited caesium might be removed from different types of building surface but it is likely that the levels would be greatly reduced by steam cleaning and by weathering.

268., The deposited fission products would constitute a source of radiation dose commitment to people living in the area. This dose commitment would be assessed by measurement of radiation levels, and based on these assessments some people might need to be evacuated for periods of weeks, months or longer. There might need to be limited evacuation of people as far as 50 km from the reactor site until the radiation had decreased sufficiently from its initial level. At distances of about 15 km, the dose rate after a few months might be 5 rem per year, but deep ploughing and cleaning up in towns would probably reduce the level, and the consequential dose commitment, to the point where it would be reasonable for people to return. At distances shorter than about 15 km the clean up problem could well be very difficult and expensive. The difficulty of estimating the number of people who might be affected is clear from our remarks in paragraph 265. A reasonable estimate is that the number of people likely to be severely inconvenienced would be about 20,000, though this number could clearly be higher or lower.

Safety considerations

269. The existence of hazards on this scale forces consideration of criteria by which their acceptability may be judged. An obvious course is to relate them to other hazards that exist, whether man-made or natural in origin. The incidence

of fatalities from various kinds of disaster in the USA is shown in Figures 13 and 14([34]). (In Britain the incidence of natural disasters is much lower but the effects of some man-made disasters, such as air crashes on people on the ground or chlorine releases, would probably be much greater because of the higher population density.) The curves are characterised by a negative slope, corresponding to the fact that the likelihood of a disaster diminishes as its scale increases. They indicate, for example, that an air crash killing several hundred people on the ground, or a dam failure killing about 10,000 people, would be likely to occur about once in every 1,000 years.

270. Such events generate public concern that is much greater, in proportion to the numbers killed, than accidents causing a few casualties, and would certainly lead to strong demands for action to prevent their recurrence. Nevertheless, at least implicitly, man-made hazards of the order indicated in Figure 13 are accepted by society in relation to the benefits conferred by the technologies that create them. If it could be ensured that the risks of reactor accidents causing fatalities were comparable with, or less than, those from these technologies, then it could reasonably be argued that the same acceptance should apply to nuclear power. The risk we would be concerned with is that arising from the nuclear programme as a whole, which will clearly increase in rough proportion to the number of reactors in operation.

271. Considerations of this kind lead to the formulation of safety criteria for reactors which relate the acceptable frequency of an accidental release of radioactivity to the magnitude of that release, as shown in Figure 15. This curve was proposed by F. R. Farmer who considered that an acceptable and realistic target for reactor design was that a single reactor should present a risk of less than 10^{-6}(i.e. one in a million) per year of a release of 10^6Ci of iodine–131. The total risk that would arise from a very substantial nuclear programme based on reactors having safety characteristics defined by this curve is shown in Figure 13. It will be seen that such a programme would present a considerably lower risk than those for the other events described. Such a higher degree of safety in relation to other hazards was considered necessary in view of the special public concern that would be generated by a radioactivity release, and also to provide a contingency against the possibility that the desired safety characteristics were not fully achieved in design. We consider this point further below.

The principles of safety analysis

272. The graphs shown in Figures 13 and 14 are to varying degrees supported by extensive data relating to past events. In contrast, the risks associated with major technological developments of which there is relatively limited experience, such as nuclear or certain chemical plant, can of necessity only be estimated. Indeed, we cannot expect to validate from actual experience of accidents that the low order of risk required from such plant has been achieved. Assurance on this point can be obtained only by the painstaking analysis of plant design from the safety viewpoint.

273. With respect to reactor safety, such analysis must seek to identify all the possible failure mechanisms by which a release of radioactivity could be brought

FIGURE 13

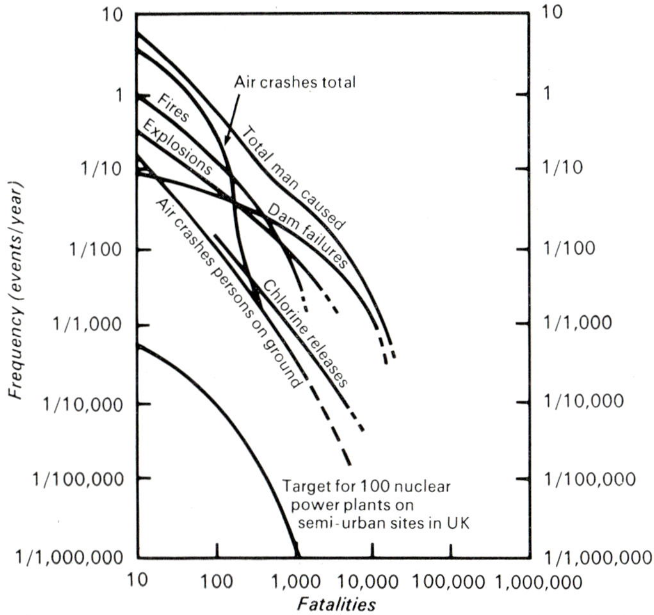

Incidence of man-made disasters (in the USA).

FIGURE 14

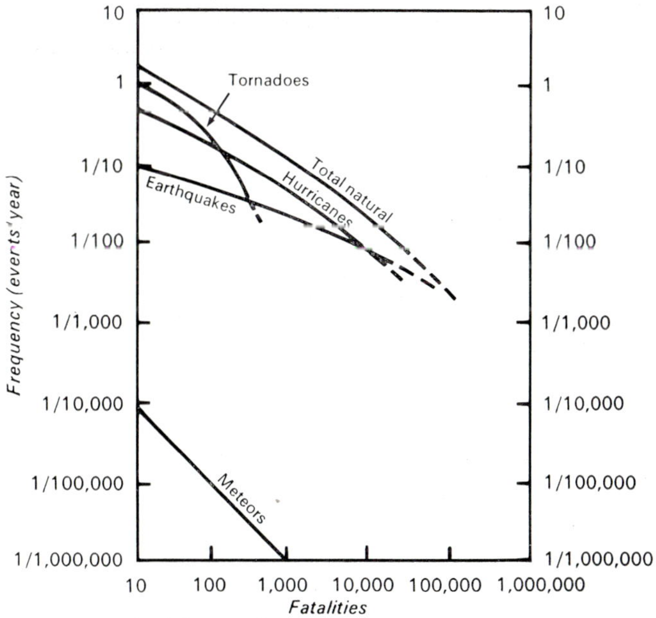

Incidence of natural disasters (in the USA).

US Nuclear Regulatory Commission.

FIGURE 15

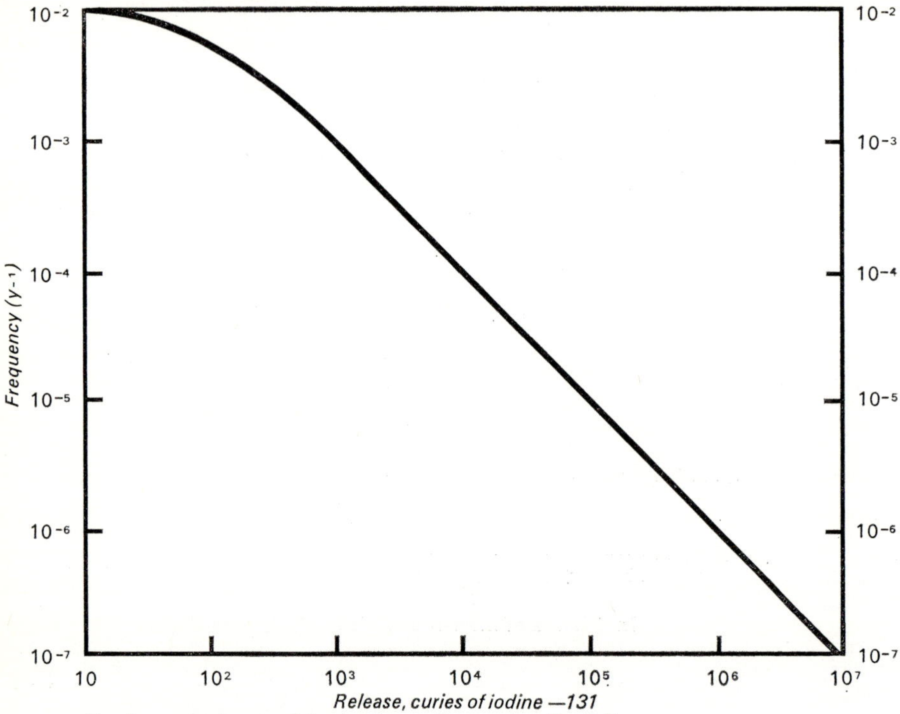

Nuclear reactor accident release frequency limit line

United Kingdom Atomic Energy Authority.

about and establish the probability of each occurring. Thus, it might be that the failure of three separate components would have to occur in sequence to cause a particular accident, and that the chance of each individual failure was known to be less than 1 per cent. The cumulative probability of the accident being caused by this route would then be less than $0{\cdot}01^3$ or one in one million. There would generally be other routes leading to the same accident and these would be represented diagrammatically by a "fault tree". The twigs would each represent a starting event, the branches the means of propagation, leading to a cumulative probability of the particular accident as represented by the trunk. Such a fault-tree, with associated probabilities of occurrence, can be drawn for any system given that the probabilities of individual component failure are known. Many of these components, such as pumps, valves and electrical relays, will already be in widespread use so that reliable data on their failure rates will be available. If there are insufficient data on particular components they will need to be tested under simulated operational conditions until their failure rates are established. Such tests will reveal not only the probability of failure but also its mode, thus directing attention to improvements that could be made. Subsequently, records should be kept of component performance in actual service. It is very important that there should be a rigorous procedure for the feedback to the design and safety authority of information on all faults that occur, though these may be small and cause no danger. Such data will increase understanding of failure mechanisms and of the reliability of forecasts of accident probabilities.

274. One aspect of safety design is the provision of "redundancy", that is, of alternative components or sub-systems so that in the event of a failure in one path the system will continue to operate. There is the possibility of "common mode" failures in which a fault occurring in one part of the system may simultaneously affect other parts and thus remove the redundancy that was thought to be present. The identification of common mode failures is an important requirement for safety analysis of the system. The analysis must also take account of the effects of human error in reactor operation and of the possibility of malevolent action. It would be an aim in design to counter such effects and the analysis can reveal how likely such counter-measures are to succeed.

275. The application of the technique of failure analysis described above in principle allows a given reactor design to be assessed against a risk criterion such as that specified in Figure 15. A fault-tree would need to be constructed for the whole system and the total probabilities attaching to a range of possible releases could then be established. If at any point the risk was greater than that allowed by the criterion, then redesign would be required to reduce the risk. That this is a practical method of proceeding is demonstrated by the Prototype Fast Reactor (PFR) at Dounreay which was designed by the AEA to fulfil Farmer's criterion. The same technique could be, and indeed is being, applied to potentially dangerous non-nuclear plant. For example, an ethylene-oxide plant has been redesigned on a probabilistic basis to a target for a major release of 10^{-5} per year.

276. The probabilistic approach to safety analysis has been developed because of the growing realisation of the need for rigorous discipline in the design of plant, whether nuclear or not, which constitutes a serious potential hazard. It provides a logical basis for the analysis and measurement of risk against specified objectives, in contrast to intuitive judgments, based on experience, which might well overlook possible failure modes in complex plant. Naturally, the technique has its limitations. It cannot be known with certainty that every possible failure mode has been foreseen in the analysis, or that the technique has been applied correctly by all the design staff involved. There may well be doubts about the long-term reliability of some components in the conditions under which they are used which cannot be entirely allayed by prior testing. Whether the safety performance is achieved in practice will depend to some degree on factors that have to be assumed in analysis, such as the quality of maintenance. There seems no doubt, however, that the technique is the best available to achieve safety in design. Whether the best is good enough in relation to nuclear hazards is perhaps a separate question. We discussed this issue in paragraphs 174–176, where we expressed the view that the hazards posed by a reactor accident were not such as to suggest that nuclear power should be abandoned on these grounds alone, but that the risks should be recognised and weighed in decisions on nuclear development.

Reactor design and construction in the UK

277. Reactors are originally developed by the AEA, who build a small prototype. Successive examples have been the Magnox reactors at Calder Hall and Chapelcross (now owned and operated by BNFL), the prototype AGR at Windscale, the SGHWR at Winfrith in Dorset, and the experimental and prototype Fast Reactors at Dounreay. They are designed, built and operated to safety standards that are laid down by the Authority's Safety and Reliability Directorate (SRD), which can report direct to the Chairman of the Authority. Thus, although the Authority is not subject to an external review in building and running its reactors and other plant, there is provided an independent internal organisation that provides an effective check on what is done.

278. The SRD have, in fact, established a high reputation for their competence and independence, and they have extended their area of operations to include non-nuclear work for outside customers. (They are also retained as consultants by a number of organisations in the nuclear industry overseas.) This has had two benefits—it has widened their experience and it has helped to bring about a better appreciation in the non-nuclear field of the techniques of safety analysis. The Health and Safety Executive (HSE) have instituted a collaborative arrangement with the SRD; in particular, the SRD are taking part in the work of the Major Hazards Committee which was set up by the HSE to examine the safety issues posed by non-nuclear plant.

279. Commercial nuclear reactors are ordered by the utilities, CEGB and SSEB, from a single nuclear reactor design and construction company, the Nuclear Power Company Ltd. (NPC). This is a wholly-owned subsidiary of the National Nuclear Corporation, in which the state has a minority interest. The NPC, which was formed from the two previous groups, The Nuclear Power Group (TNPG) and British Nuclear Design and Construction (BNDC), has been constructing the Hinkley Point "B" and Hunterston "B" AGRs, and is building the AGRs at Heysham and Hartlepool. The reactors are designed and built to the requirements of the utilities who in turn have to satisfy the requirements of the Nuclear Installations Inspectorate (NII) in order to be granted a site licence. Such a licence is given on the basis of the design, to enable construction to begin, and subsequently a further licence is granted to enable the reactor to be loaded with nuclear fuel and brought to criticality.

The Nuclear Installations Inspectorate (NII)

280. The NII is a body of some 90 engineers and scientists which, since 1975, has been part of the Health and Safety Executive (HSE). The Executive reports to the Health and Safety Commission, which formally issues site licences for the whole of Great Britain under powers in the Health and Safety at Work &c Act 1974*. Formerly the NII was part of the Department of Energy and its

*Parliamentary responsibility for nuclear safety remains with the Secretary of State for Energy in respect of England and Wales, and with the Secretary of State for Scotland north of the border. In Northern Ireland, nuclear sites would be the responsibility of the NI Department of Commerce, which comes under the Secretary of State for Northern Ireland, but the technical advice would again come from the Nuclear Installations Inspectorate under the 1965 Act, whose provisions in respect of Northern Ireland were not repealed by the Health and Safety at Work &c Act.

function as defined by the Nuclear Installations Act 1965 was to provide technical advice to the Secretaries of State for Energy (for installations in England and Wales) or for Scotland, who then issued the site licences.

281. Nuclear site licences are very lengthy documents, which specify in considerable detail the conditions under which the plant may be operated. They may be altered or revoked at will, and the operator can be ordered to close down a reactor or reduce its rating at any time. These reserve powers are occasionally used; thus, the CEGB were required some years ago to institute further checks before a reactor at Bradwell was permitted to restart after a shut-down for maintenance. Although the NII can prosecute operators for infractions, they have not needed to do so.

282. The NII thus have very substantial powers to control how nuclear reactors (and other installations such as those used for uranium enrichment, fuel reprocessing and radioisotope production) are designed, built and operated. At the major sites, there is an Inspector present several days a month on average, and within the Inspectorate there is the technical capability to check reactor design in considerable detail. The NII do not have a research laboratory of their own, but they can and do commission such research as they need from outside bodies, for example university departments. We do not doubt the technical competence of the Inspectorate or the thoroughness with which they carry out their work. However, the discussions we have had on reactor safety with several authorities have left us with some doubts about whether the criteria adopted by the NII in establishing reactor safety are soundly based and whether their functions are correctly defined. We have not investigated these matters deeply and we express our views in the following paragraphs with the object of assisting their further consideration by the authorities concerned.

The maximum credible accident

283. The concept of "maximum credible accident" (mca) is basic to the licensing of reactors in this country and others, including the USA. For a given design of reactor, failure mechanisms are postulated which the licensing authority consider would result in the worst possible accident that could occur in practice. It is then a requirement of licensing that the reactor should be so designed that such an accident, the mca, would be safely accommodated so that any consequential release of radioactivity would not cause significant harm to the public.

284. Intuitively the concept presents difficulties, for it raises the question of the means by which it is to be decided whether theoretically possibly failure modes are to be deemed credible or incredible. Unfortunately, it is a common experience in life that incredible things do happen, no matter how surprising that they happen to us, and when they do some new factor or combination of factors not previously allowed for is usually to blame. In fact, it is admitted that a chain of events can be imagined that would result in an accident with consequences more severe than those of the mca, but the chance of this taking place is regarded as too small to be taken into account.

285. The fact that a reactor has been designed so that the mca would cause no harm conveys the impression that absolute safety has been achieved, though such an impression would be illusory. There is a conflict between the kind of assurance about safety that the public would expect to receive and governments like to give, and the kind that engineers can supply; that is, between such statements as "this reactor is completely safe and presents no danger to the public" and "this reactor is so designed that the chance of an accident that would cause serious harm is considered to be less than (say) 10^{-6} per year". The public are not generally so acquainted with probabilities as to understand the significance of a statement of the latter kind; it might increase rather than allay anxiety. We can see, therefore, that the mca concept may be useful as a means of assuring the public about reactor safety.

286. We are more concerned about the use of the concept as a criterion for safety design within the industry. We described earlier the technique of fault-tree analysis which requires the disciplined investigation of the probabilities of all failure modes and, in principle, allows a design to be confirmed against specified risk objectives. It is clear that probability analysis must to some degree already be used in applying the mca concept, for the determination of what is incredible or otherwise must rest on judgment of the likelihood of failures, but the systematic use of the technique is not a condition of licensing. It is possible, therefore, that because attention is focussed on certain specific failures that would lead to the mca, there will not be disciplined analysis of other modes of failure which, though less likely to occur, could have much more serious consequences, and whose probabilities of occurrence it should be an object of design to reduce to known and acceptable levels. This may mean that resources are not used in the optimum way in relation to the spectrum of risks that actually exists.

Method of working of the NII

287. We have noted that reactors are designed to meet the specifications of the utilities, who then have to satisfy NII requirements in order to obtain the site licence which will enable construction to begin. There seems to be rather little contact between the Inspectorate and the designers during the design stage. The Inspectorate is not formally involved until the complete design has been prepared and submitted to them. The design is analysed in considerable detail and the Inspectorate then indicate any requirements they may have for changes as a condition of licensing.

288. The view has been expressed to us, and we find it convincing, that it is impossible for an inspecting body adequately to ensure safety by analysing the completed design of a large, complex plant. Safety is built into the design as it progresses and the primary task of the inspecting authority should be to ensure that the proper analysis techniques and disciplines are used at every stage of the design process to achieve clearly defined safety objectives. The second aspect of the authority's work is the routine checking that the plant is built, operated and maintained in accordance with design intentions. It will involve such matters as quality control and operating procedures and will require repeated

site visits and careful inspection of what is being done. There are thus two levels of inspection which call for different experience and qualifications in the staff concerned.

289. Strong support for the probabilistic approach to safety in reactor design was given in their evidence by the NPC who, like the SRD, regarded it as providing a valuable discipline within which the designer could work. We were interested to learn that this approach has in fact been used for a commercial reactor design, namely the AGR at Hartlepool. The customer, CEGB, gave the BNDC a contract to undertake a probability assessment of the design. It was found that the previous *ad hoc* rules were inadequate in some areas and unnecessarily stringent in others, but the main outcome was a greater assurance that potential faults and fault sequences had not been overlooked. We understand that the CEGB, who as operators would bear the immediate responsibility for safety of their reactors, were also impressed by the advantages of the probability technique.

290. We have therefore reached the conclusion that there is a need to review the criteria and methods of working of the NII and we recommend that this should be undertaken by the authorities concerned. Issues of the kind discussed above are likely to arise in a wider context in the work of the HSE Major Hazards Committee, which is concerned with non-nuclear plant of all kinds. In contrast to reactors, complex chemical plants, which may pose hazards at least as great in some respects, are not at present subject to safety analysis by any external body (unless explosives are involved). It appears inevitable that a system of licensing will need to be adopted for some non-nuclear installations. The same disciplines and safety techniques are called for in the design of potentially dangerous non-nuclear plant as for reactors, and there may well be advantages in combining responsibilities for ensuring their use in both fields within a single inspectorate.

The provision of advice to Government on nuclear safety

291. We have noted that the NII are now within the HSE rather than, as formerly, the Department of Energy, which is the sponsoring department for nuclear power. We welcome this change; we prefer to see responsibilities for the promotion and regulation of an industry within separate organisations, since conflicts of interest are then likely to be more fully and openly resolved.

292. We are concerned about the need for a source of independent, expert advice to the Government on technical matters that are relevant to policy decisions on major and hazardous technological developments, whether nuclear or otherwise. Such a decision for example would be whether to embark on the design and construction of the first commercial scale fast breeder reactor. So far as nuclear matters are concerned it appears that at present the main source of authoritative advice is the AEA, a body that is inherently committed to nuclear development. The advice would stem from the SRD, and we have already commented on their competence and on the degree of independence they enjoy with the Authority. Still, on matters of this consequence it is in principle

desirable that advice should be available from a manifestly independent body. There was previously a Nuclear Safety Advisory Committee which had this advisory function; this Committee has been dissolved, though we hope and believe that it is to be reconstituted by the HSE. It appears to us that the HSE have a responsibility to give such advice and we think that they should develop the capability to do so. We acknowledge a basic problem, that the ability to advise rests on technical competence which is difficult to acquire and sustain outside the industry concerned. We suggest that it might be advantageous, at least for nuclear matters, for a few HSE staff to be appointed who would normally reside with the SRD and who would thus be in a position to acquire direct experience and understanding of developments on safety questions. They would be able to make use of expertise within the SRD and to represent HSE requirements with regard to any work undertaken on their behalf.

Siting policy for reactors

293. The hazards posed by nuclear reactors were recognised when they were first developed on a commercial scale in this country and they were therefore sited in areas remote from large towns. Coastal siting was dictated by cooling water requirements* since the UK has no rivers of adequate size to cool a really large station and cooling towers are expensive and unsightly. (In other countries, where nuclear power stations are sited on major rivers, the hot water discharges have been a cause of much complaint and a focus for opposition to nuclear power.) But sites for nuclear power stations must meet stringent criteria, and coastal sites that meet these criteria and are not in national parks or areas thought to be of outstanding natural beauty are few. Where nuclear stations are sited in such areas we think that consideration should be given to preserving public access to the coastline. Coastal siting raises the additional problem that long transmission lines are generally needed to take the electricity to the load centres. These are also the cause of complaint on grounds of amenity, and in fact when Public Inquiries are held on a proposed power station, such issues tend to overshadow those of reactor safety. Indeed, safety as such has hardly been an issue and the Torness Public Inquiry took place before it had been decided whether an SGHWR or a PWR was to be built there.

294. During the time when the shortage of suitable sites was becoming apparent, modifications were being made to nuclear reactor designs that gave greater confidence about safety, in particular the change from a steel pressure vessel with external boilers to a pre-stressed concrete one with boilers inside that was adopted for the later Magnox stations and the AGRs. Accordingly, a review of siting policy took place under the Nuclear Safety Advisory Committee. The new policy was announced by the then Minister of Power in February 1968. A more relaxed criterion was adopted for the numbers of inhabitants permitted within 30 ° sectors from the station at different radii and for the total

*Nuclear power stations generally require twice as much cooling water as fossil-fuelled ones for the same electrical output. The latter disperse about 15 per cent of the heat from combustion up the chimney and they typically have a significantly higher thermal efficiency as well. The cooling water requirement is less for AGRs and FBRs than for Magnox and LWRs since the former have a higher thermal efficiency.

numbers within given distances. As a consequence, the sites at Heysham and Hartlepool (which is only 5 miles from Middlesbrough) were deemed acceptable, and chosen as the first two "near-urban" sites.

295. It is Government policy that future commercial reactors should all be acceptable in principle for "near-urban" siting, although it is intended that the first few reactors of a new type should be sited remotely so as to gain practical experience of their characteristics. The safety of the public is considered to derive more from high standards in the design, construction and operation of nuclear power stations than from remote siting. We agree, and would go further. Because of our views on the desirability of using the waste heat from power stations for district heating (see paragraph 491), we should wish to see nuclear stations developed that could be sited sufficiently close for this purpose to areas where a large enough heat load exists; this would dictate siting within about 30 km of the urban areas involved. The need for transmission cables would also be reduced, and hence their adverse effects on amenity. We acknowledge, however, that urban siting would present some conflict with security considerations (see paragraph 309) and this aspect would have to be considered in deciding policy.

296. There are some inconsistencies in the evidence we have received on siting policy. According to the NII, risk factors for sites ranging from within London to remote areas vary by three orders of magnitude, that is, a thousand-fold. But according to the SRD, the risk varies over this range only by a factor of about thirty. It appears that the difference arises because of the different approaches of the two bodies to risk appraisal. The consequences of the mca would be limited to an area fairly close to the reactor site and hence would be expected to vary greatly between city and remote area siting. On the other hand, if much more serious, though much less likely, accidents are taken into account a much greater area will be involved, and heavily populated areas could be affected even though the reactor is remotely sited. A lower range of risk might thus be expected. We note that the NII site criteria include limits on the number of inhabitants up to 20 miles even though the mca (for an AGR) would give radiation doses outside the station site of only one-fifth of the emergency reference level. There appears to be a contradiction here in that releases higher than those that would arise from the mca are regarded as sufficiently probable to justify allowance for them in siting policy. Our conclusion is that there is a need for siting policy to be reviewed.

Fast reactor safety

297. In paragraphs 100, 101 and 115 we briefly described the differences between thermal and fast reactors in their mode of operation and from the safety viewpoint. Because of the importance of the FBR in relation to nuclear development, we conclude this chapter with some further comments on the relative safety aspects.

298. For reactors of either type, safety depends on achieving the required high degree of reliability in three main areas: the control systems, such as those

used for shut-down; the so-called "engineered safeguards", such as emergency core cooling systems; and structural integrity. In each of these areas the liquid metal-cooled FBR has advantages and disadvantages compared with thermal reactors.

299. On thermal reactors the most difficult safety problems arise through the use of pressurised coolants. A light water reactor, for example, contains water at very high pressure and at several hundred degrees; sudden depressurisation would release very large amounts of energy (whose effects would be hard to predict) and there would be a sudden drop in cooling ability as water around the fuel elements flashed into steam. In water cooled reactors, therefore, emergency core cooling systems (ECCS) have to be provided. There are difficulties in ensuring the effectiveness of such systems; thus, if the fuel elements are distorted by the heat the adequacy of water circulation by the system may be in doubt. Until recently, the postulated effectiveness of ECCS has depended on computer calculations, but full-scale experiments are now taking place in Idaho.

300. The FBR has the advantage that the liquid sodium coolant is not pressurised and has a high heat capacity. Whereas in thermal reactors the coolant has to be kept in motion mechanically by pumps, in the FBR, even if the pumps failed, natural convection in the sodium would maintain a high rate of heat removal and would allow a time of the order of an hour before auxiliary cooling would have to be brought into use. This would be supplied by one of several alternative systems provided in the design. On the other hand, there are problems in handling sodium on a large scale, since it constitutes a major fire hazard. Experience suggests that these problems can be solved and even quite large sodium fires have been satisfactorily extinguished. It should be noted that the presence of sodium precludes the ultimate use of water to extinguish a reactor fire; this method was used eventually to extinguish the fire in the Windscale reactor which occurred in 1957.

301. The use of a low pressure system in the FBR avoids the problems of high-pressure containment which make it difficult to ensure the structural integrity of thermal reactors. In LWRs, for example, this depends on thick-walled steel pressure vessels where the technology of welding has been taken beyond previous limits in respect of both size and thickness; safety depends on interpretation of research work rather than on generations of experience. There are difficulties in the detection of cracks in such thick structures and in repair, which must be done *in situ*. Reinforced concrete pressure vessels, such as are used for AGRs and the later Magnox reactors, are thought much less likely to suffer catastrophic failure, though the possible mode and consequences of leakage from the steel liner must be considered and the effects of temperature changes and of ageing.

302. With any reactor employing a liquid coolant there is the risk that if a substantial amount of the fuel should melt it could react explosively with the coolant, causing the latter to vaporise. The phenomenon takes place from time to time in steelworks when molten steel is poured into a casting mould or a

ladle containing some residual dampness([32]); the resultant instant vaporisation of the water causes an explosion which is sometimes catastrophic. In a fast reactor the transfer of heat would be hampered by the relatively low thermal conductivity of molten uranium and plutonium oxide and the relatively much greater amounts of sodium that would surround the molten fuel. Moreover, a reactor that has been running for a little while will contain significant amounts of gaseous fission products in the fuel (notably krypton and xenon) and these would tend to form a blanket to prevent intimate contact between the fuel and the sodium. Still, the possibility exists, and is particularly serious for the FBR because of the risk that the explosive forces generated might so change the core geometry as to lead to a prompt critical situation. There would then be a cumulative increase in reactivity which would be limited only by the core blowing itself apart, that is, by "disassembly" of the core, and its vaporisation and dispersal. No mechanism can be foreseen by which this could happen in gas-cooled Magnox and AGR reactors though a possible mechanism can be postulated for water cooled reactors. In such reactors the increase in reactivity that could be caused by the rapid removal of a control rod might be sufficient to bring about prompt criticality; this would imply that the rod was blown out of the core but measures to prevent this possibility are included in the design. Thus the FBR is not unique in posing the hazard that a prompt critical condition could arise, but the event would be much less improbable for the FBR where a greater range of circumstances might lead to its occurrence.

303. Earlier in this chapter we discussed the effects of a major release of the gaseous and volatile components of the fission products from a thermal reactor. The inventory of fission products for an FBR is likely to be similar for a reactor of about the same output capacity, but the core will contain several times as much plutonium, and an even greater proportion of the higher actinides, americium and curium. If fuel were to melt inside the sodium, most of the fission products would be retained within the sodium because of chemical reactions. If some burning sodium was expelled from the reactor, the fume would carry any associated fission products. It is considered that for most such accidents that can be envisaged the fission products would be retained within the very strong structure within which the reactor is housed. However, the possibility of gross vaporisation of the fuel, and the subsequent escape of some proportion of the inventory to the atmosphere, overrides all other considerations. This would allow not only gaseous and volatile fission products to escape but also those that are normally non-volatile. It has been estimated that a 10 per cent release from a large FBR would have consequences one to two orders of magnitude worse than those we have described for a thermal reactor. The major effect would be due to the inhalation of the non-volatile fission products, leading to the development of lung cancers, by people downwind of the release. Ground contamination would be caused by plutonium as well as caesium.

304. The consequences of such accidents would be so severe that the FBR could scarcely be contemplated on an extensive commercial basis unless it is established with a very high degree of confidence that reactivity accidents can be controlled so as to prevent gross fuel vaporisation. It is generally con-

sidered that present research programmes will lead to substantially increased knowledge of these effects within the next five years. Work is proceeding in many countries as it is recognised that the problems are of fundamental importance if fast reactors are to be widely used. One important source of knowledge is operational experience from prototypes such as the PFR at Dounreay. The behaviour of fuel under irradiation and under a variety of simulated emergency conditions can be studied, although the latter conflict somewhat with the need to establish reliability as a normally operating power plant. The incidence of local boiling of the sodium coolant must be detected quickly, and the means likely to be used is to monitor the system for the noise that such boiling would cause. There are, however, other safety issues that can be resolved only in the design process itself, and thus the design and construction of a first FBR of commercial scale is an important step in assessing whether the required level of safety can be achieved.

CHAPTER VII

SECURITY AND THE SAFEGUARDING OF PLUTONIUM

Introduction

305. Much of the concern that is felt about the environmental impact of nuclear power relates not to the effects arising from normal operations but to those that might be created by illicit activities directed towards nuclear installations or materials. The safeguarding of society against such activities raises issues which are in some ways beyond our remit, but we felt compelled to give some consideration to them for, had we not done so, our study would justly have been regarded as seriously incomplete. We are not experts in security matters and we do not know the details of the security arrangements already in operation or planned. We look at these problems from the standpoint of the ordinary citizen; this is an important viewpoint, for much depends on the reaction of the ordinary citizen to the restrictions on his freedom, real or apparent, which might result from security measures. We thought we should try to establish whether the general approach being taken to security measures is commensurate with the risks as we see them. Accordingly, during our study we have discussed security aspects in broad terms with the bodies concerned with nuclear development and operations, and arrangements were made for some of our number to meet the Lord Privy Seal and senior government officials to explore official thinking on these issues.

306. We drew attention briefly in Chapter IV to the security issues posed by nuclear development. There are three different aspects. One is the risk of sabotage of nuclear installations which could cause very substantial damage and perhaps lead to the escape of radioactivity. The risk of damage, of course, exists with other large industrial plants and in some cases this could lead to hazardous releases of chemicals. The second problem is that of the diversion of plutonium and the possibility that it could be made into a bomb or dispersed, accidentally or deliberately, into the environment. The opportunities for diversion will inevitably become much greater as increasing reliance is placed on the use of plutonium in fast reactors, in what some have termed the "plutonium economy". The third aspect is whether the security arrangements that will become necessary to protect society against these risks will eventually pose a serious threat to civil liberties. This is seen as a major issue in the USA.

307. It is our purpose in preparing this Report to help the public to understand the issues involved in decisions on the development of nuclear power. With respect to security matters we face a dilemma. There is the danger that in seeking to inform the public about the nature and extent of the hazards from illicit activity in the nuclear field, we may help to stimulate and direct such activity. On the other hand, if security issues are a determinant in crucial

decisions on nuclear development, then it is important that they should be widely understood and discussed. The difficulty is one that has been discussed by others writing on the subject. A view that has been expressed, and which we find persuasive, is that if the risks involved to society were temporary then the dangers of inspiring illicit activity would justify withholding information from the public. But that if, as applies to nuclear power, the security risks are long-term and likely to become increasingly severe as development progresses, then information that would enable the public to reach an informed view on the issues should be made available before the major decisions are made so that it can be fully taken into account.

308. In any case, there is already so much published information on the potential of nuclear threats, and on technical matters that would be relevant to putting these threats into practice, that we cannot believe that the analysis we present in this chapter will materially affect the risks. To suppose otherwise would indeed be to imply that the dangers are already so real that a warning is needed. We are satisfied that in this country at least, with the present state of nuclear development and with the security measures that are now in force or are being introduced, the risks to society from illicit activities are small. The main concern lies with a future in which there could be substantial growth in nuclear power and a move into the "plutonium economy". If these problems are to be publicly discussed at all, as we believe they should be, then the sooner the better.

The threat of sabotage

309. We have explained that there are very large accumulations of radioactivity, in the form of highly active fission products, within reactor cores and in storage tanks at a fuel reprocessing plant such as Windscale. In Chapter VI (paragraphs 266–8) we have described the effects of a major release of radioactivity. Elaborate precautions are taken to guard against accidental release of these large inventories, and they are additionally protected by massive containment structures (several metres of reinforced concrete) which are needed in any event to provide a biological shield for the operating staff. It would, therefore, certainly be difficult for terrorists or saboteurs to bring about a catastrophic failure and consequent release of radioactivity. But if they had sufficient knowledge of the plant design the possibility of such an event cannot be ruled out, and it is not beyond the bounds of credibility that a sufficiently intelligent and determined group could acquire this knowledge. A situation in which a terrorist group had taken control of a reactor facility, and was able to provide plausible evidence that they had this knowledge, would be a very difficult one for a government.

310. An important aspect of the sabotage threat is the economic damage that would be caused by disruption of the plant. Nuclear power stations are exceedingly expensive, costing hundreds of millions of pounds, and as potential targets for a terrorist group would be spectacularly newsworthy. Even if the plant were subsequently repaired, the loss of production could be very large.

For example, the fire at the Brown's Ferry reactors in the USA (see paragraph175) is estimated to have caused an economic loss of about £50 million.

311. There would be a particularly serious threat if in the future we came to depend upon plutonium as our main source of energy. Because of its high value, it is likely that stocks would be kept low, and a fast turn-around assured by the reprocessing plant. If there were only a few such plants, as seems likely, and one were put out of action as a result of sabotage (or even an accident) leading to serious radioactive contamination, the effects could last for years and the whole country could be very seriously affected by failing supplies of electricity. To allow a situation where a single terrorist attack could put the economic life of the country at risk would be a dangerous and, we would consider, an irresponsible policy.

312. The cost of security measures that might become necessary to protect nuclear installations is also an important factor. In the UK there has been no protest from the nuclear industry that improved security would impose an excessive financial burden, but this has occurred in the USA. However we were informed by the utilities that two proposals for increased security, namely the grouping of plutonium production plant and fast reactors onto a few integrated sites or "nuclear parks", and the provision of a sufficient armed force at a power station to prevent it being taken over by terrorists, would be uneconomic. Whereas the dangers to the public of a take-over of a Magnox station are probably not very great, the situation with a large FBR station would be quite different. There would, for example, be a stock of new fuel sub-assemblies, which could be handled quite safely; each one would contain a substantial quantity of plutonium, though the element would be in mixed oxide form with uranium and its separation would require difficult chemical reprocessing.

313. Another example where the need to balance cost and security requirements may arise is in the provision of railheads at nuclear power stations. Were new fast reactor fuel elements invariably to be transported by rail from factory to reactor site, the opportunity for diversion would be much less. (It is assumed that they are conveyed in some very heavy and secure container as a precaution against accidents.) It has been argued before us that the provision of a railhead for nuclear power stations would be desirable from the point of view of movement of the large containers used for the transport of irradiated fuel. Although this proposition was advanced on grounds of increased safety, it would also have advantages for security. Certainly the containers (see Plate 7) are massive and could not easily be breached, but their movements will be very numerous in the future. While they seem unlikely to be attractive targets for attack, the effects of explosive charges could possibly be serious and we hope that this aspect has been adequately investigated.

314. We have given some thought to the possible effects of war so far as nuclear installations are concerned; these installations, providing vital energy supplies, would be prime targets. In a nuclear war the effects of attack on nuclear installations would be one part of a general catastrophe, but an attack with

conventional weapons leading to the release of radioactivity would produce some of the effects of nuclear weapons. The quantities of fission products that could be released are vast and they would not be carried up into the stratosphere.

315. The effects of war, even of "conventional" war, are inevitably horrifying, but if these effects could be magnified by attack on nuclear installations, then this is a major factor to consider when deciding whether, or to what extent, to use nuclear power. This threat also exists, and should likewise be weighed, in the non-nuclear field. The vast increase in the chemical process industry over the last few decades has created many industrial plants where dangerous substances are used or stored and where the consequences of damage from armed attack could be extremely serious. The unique aspect of nuclear installations is that the effects of the radioactive contamination that could be caused are so long lasting. If nuclear power could have been developed earlier, and had it been in widespread use at the time of the last war, it is likely that some areas of central Europe would still be uninhabitable because of ground contamination by caesium.

The diversion of fissile materials

316. We are concerned mainly with plutonium; as we have previously explained, the mainstream of current nuclear development leads to the widespread use of breeder reactors which produce plutonium for reactor fuel. It should be noted that an alternative line of development, based on the thorium cycle, would avoid the production of plutonium, but it would require the use of highly enriched uranium and would lead to the production of the isotope uranium–233. Uranium in both of these forms can provide the core material for a fission bomb so that one of the principal dangers of plutonium would not thereby be avoided.

317. The value of plutonium and its potential for nuclear threat would seem to provide ample motive for theft. Some people believe that as the quantity of plutonium in existence increases, together with the numbers of people concerned with its use and aware of its potentialities, a market for illicit materials is bound to develop, and theft to supply this market is inevitable. We emphasise again that we are considering a situation that could develop in the future. At the present time plutonium is stored in fairly small quantities at a few sites where appropriate security arrangements can be adopted. On the AEA's projection of nuclear growth in the UK (see Table 14, paragraph 468) however, the amount of plutonium produced by the year 2000 would be a total of 250 tonnes growing to more than ten times this amount by 2030.

318. The element would need to be transported on a considerable scale. The risk of its diversion would be particularly great if the element was in a form in which, given a knowledge of necessary precautions, it could be handled without danger. The plutonium in irradiated fuel can be extracted only by complex chemical reprocessing and the intense radioactivity is such that these operations could certainly not be undertaken illicitly. The plutonium oxide in new fast reactor fuel elements could in principle be separated out, although it is mixed

with uranium oxide and there would still be a need for chemical processing. Such a task would become appreciably harder (and more dangerous for the thief) if the fuel elements were to be briefly irradiated in a reactor before shipment in order to make them highly γ-radioactive. Within the UK at present (and, it seems, for the foreseeable future) fuel reprocessing and the manufacture of mixed oxide fuel elements will be undertaken at the same site so that significant movements of plutonium other than in fuel elements will be unnecessary. These fuel elements would need, and would receive, protection during transport and storage prior to loading into a reactor.

319. The utility of the plutonium to a potential terrorist would be particularly great if it were shipped as a pure compound, oxide or nitrate. The latter form is especially hazardous from the point of view of dispersion. In other countries notably the USA, pure plutonium compounds are transported because the various nuclear facilities are dispersed. The UK could be involved in such shipments with their increased attraction for potential thieves, were we to return plutonium removed from fuel reprocessed for foreign customers in this form. We consider that this is an unnecessary risk, and that as a matter of policy, such plutonium should be returned only in the form of mixed fuel elements designed to suit an existing power reactor.

320. Plutonium could be illicitly obtained either by the persistent theft of small quantities by personnel within reprocessing and fabrication plants or by terrorist action against these plants or during transport. The risk of theft within the plants is minimised by personnel checks and by strict security arrangements for access to the material. Elaborate accounting procedures are used to keep a check on the incoming and outgoing quantities involved in any process, and hence to detect any losses and to deter illicit removal. However, there are inevitable uncertainties in materials accounting especially at the initial stages of plutonium separation* and as the amount of plutonium in circulation increases, quantities that would be significant for illicit purposes might be removed without detection. The danger is not so much that such quantities would form a progressively smaller fraction of annual throughput, because the amounts of the element in a section of plant to which checks are applied will always be limited in order to avoid the risk of a criticality accident. It is rather that as plutonium comes to be regarded as a routine item of commerce, there may be a slackening in the vigour with which safeguards are applied.

321. The risk of loss through direct terrorist action depends on the detailed security arrangements adopted for the storage or transport of plutonium in relation to the organisation, expertise and ruthlessness of potential terrorist groups. Some people concerned with this matter have postulated groups consisting of about a dozen disciplined and well-armed terrorists, possibly assisted by "inside" information and support; the latter might be obtained by threat against personnel or their families. If such groups have indeed to be assumed then the security arrangements required in the plutonium economy would be extensive and their implications for society should be considered.

---

*See footnote to paragraph 132.

322. The hazards of plutonium arise from its extreme radiotoxicity and from its explosive properties. The dispersion of a small amount of plutonium into the atmosphere with conventional explosives would pose a very serious radiological hazard since an individual dose of only a few milligrams is sufficient, if inhaled, to cause massive fibrosis of the lungs and death within a few years. Much smaller quantities can cause lung cancer after a latent period of perhaps 20 years. In fact it appears highly probable that only a small proportion of the plutonium would be disseminated in the particle size range that is respirable, but apart from the actual threat the psychological effect of a dispersion device would be very great. If such a device were to be detonated in a populous area, very expensive decontamination operations would certainly be needed.

323. The construction of a nuclear bomb by a terrorist group would certainly present considerable difficulties and dangers to those attempting it. The equipment required would not be significantly more elaborate than that already used by criminal groups engaged in the illicit manufacture of heroin, but great care would need to be taken in the handling of dangerous materials and in avoiding accidental criticality. A substantial knowledge would be needed of the physical and chemical processes involved, of the properties of high explosives and of the principles of bomb construction. We have been impressed and disturbed by the extent to which information on all these topics is now available in open technical literature.

324. There seems no reason to doubt that a sufficiently determined group with the necessary expertise could construct a very crude bomb which might explode with the force of a few tonnes of TNT. The amount of plutonium required could easily be carried by hand. Though extremely inefficient in nuclear terms such a device would still cause much damage and would create immediate radiation which would be lethal over a range of several hundred metres as well as dispersing radioactive material over a wide area. More doubt attaches to whether an illicit group could construct a weapon with a much greater yield, say 100 tonnes of TNT or more. There is some dispute about this possibility. From the discussions we have had we have formed the impression that the British authorities are less persuaded than those in the USA about the credibility of the construction of such a weapon. We felt it necessary to settle the matter in our own minds and we therefore consulted eminent physicists both in the UK and the USA who are expert in the subject. Their judgment was that the construction of a bomb that would give such a yield was indeed possible, though the actual yield would be very uncertain, for it would be as much a matter of luck as of good judgment.

325. We have concluded therefore that it is entirely credible that plutonium in the requisite amounts could be made into a crude but very effective weapon that would be transportable in a small vehicle. The threat to explode such a weapon unless certain conditions were met would constitute nuclear blackmail, and would present any government with an appalling dilemma. We are by no means convinced that the British government has realised the full implications of this issue.

International safeguards

326. We conclude our discussion of the plutonium diversion problem with a brief description of the system of international safeguards. Security arrangements are no stronger than their weakest component, for plutonium stolen in one country could be put to illicit use in another. A number of observers have suggested that large parts of the world are inherently unsafe for nuclear power([33]) from this and other viewpoints; we were interested to learn that the CEGB agreed with this view. A system of international safeguards to prevent the diversion of fissile materials has been set up under the International Atomic Energy Agency (IAEA) in Vienna. These safeguards are mandatory on all non-nuclear weapons nations party to the Nuclear Non-Proliferation Treaty (NPT); we referred to general aspects of the proliferation problem in paragraphs 165-7 (Chapter IV). The safeguards have also been accepted on a bilateral basis by a number of other nations which do not subscribe to the treaty but wish to benefit from peaceful uses of nuclear power promoted by the IAEA.†

327. Under these safeguards, a national system of accounting for fissile material has to be devised and implemented, and duplicate copies of the inventories of each Materials Balance Area* (MBA) and of additions to and subtractions from them, have to be submitted regularly to the IAEA. The Agency employ inspectors who check the system at intervals, and ensure that all is in order. Penalties for infractions could include the return of fissile material and the loss of technical assistance by the Agency, but it is likely that international pressures would be applied if it were suspected that there was a significant amount of Material Unaccounted For (MUF) that had been diverted at the orders of the national Government.

328. The IAEA safeguards are carefully designed to provide essential checks without infringing national susceptibilities but it seems probable that there will be difficulty in preserving the effectiveness of the system given the tremendous expansion of nuclear power that is projected for the future, which would imply a great increase in the work of the inspectorate. The extent of this increase may be gauged from the fact that under the formula relating the maximum number of inspections to the throughput of plutonium at a plant, Windscale could receive the attentions full-time of nearly six inspectors at present, and twenty five inspectors by the end of the century. There are at present only 74 inspectors to cover the whole world.

329. In practice Windscale, being in a nuclear-weapons state, does not receive full inspections, and even after the voluntary offer by the British Government (in parallel with one by the USA Government) to place all our civil nuclear power activities under IAEA/NPT Safeguards has been implemented (probably late in 1977), it is likely that the IAEA will still concentrate their attention on nuclear facilities in non-nuclear weapons states. In fact Windscale has been

†Safeguards under the NPT differ somewhat from non-NPT ones and are generally more rigorous. Attention is focussed on the accountancy of fissile material rather than on the security of the plant.

*This is an area of operation which can be treated as independent for accounting purposes. There are about six MBAs in the Windscale works.

subject to international safeguards inspection since 1973, when the UK joined the European Communities and consequently EURATOM. EURATOM operates a system of safeguards similar to but not identical with those of the IAEA; this is regarded by the IAEA as a "national" system, requiring some further independent checking by IAEA inspectors. The EURATOM inspectors visit the Windscale site about once a month, and details of the movements of fissile materials are forwarded to EURATOM headquarters at this frequency.

Effects on civil liberties

330. The presence of international inspectors and a system whereby the movements of fissile material are carefully monitored are both developments that we welcome. In fact the nominal loss of national sovereignty involved in having our civil nuclear power activities checked by foreign inspectors is the inevitable price of reciprocal arrangements being agreed in other countries where we might have more cause for concern about the chances of diversion of fissile material. But fears have been expressed that a safeguards system, to be really effective in deterring theft, in recovering stolen plutonium and in dealing with threats of terrorism or blackmail, may have to take on a more active role in the population at large. We referred to these concerns in Chapter IV: they have been most apparent in the United States where there is a Bill of Rights guaranteeing individuals against the excessive power of the state.

331. There are three main grounds for concern. The first is with the rights of employees who work in a plutonium factory. They will require to be screened before being employed, as they already are, and they may be subject to unusual surveillance during the course of their employment. Security screening is of course usual in some other areas of employment; though the need for it may be accepted by prospective employees, its more disagreeable aspects, especially the approaches made to friends, are resented by many. Security requirements are relevant to the question whether private, non-state, organisations and employees should be involved with the handling of plutonium and other fissile materials in the nuclear fuel cycle. This happens in the USA, where there have been quite a large number of prosecutions (and fines) for breaches of security. In the UK, there exists at present a monopoly exercised by BNFL, which, although nominally a private limited company, is in fact owned by the state, and required to observe the same security standards as the AEA, which is additionally charged with ensuring compliance. (The two organisations use the same security arm, the AEA constabulary). The Secretary of State for Energy retains the residual power to permit other organisations to hold or make fissile material, and we do not think that such permission should be granted to non-State organisations.

332. The second issue is over the secret surveillance of members of the public and possibly of employees who may make "undesirable" contacts. The activities might include the use of informers, infiltrators, wiretapping, checking on bank accounts and the opening of mail, and they would be practised on members or suspected members of extremist or terrorist groups or agents of foreign powers who it was thought might plan an attack on, or theft from, a

plutonium plant. We regard such activities as highly likely, and indeed inevitable. No doubt these methods are already applied to certain small groups that are regarded as dangerous, so that their use in relation to the plutonium threat would constitute nothing new in principle. It is a matter of degree, and the real question is the extent of the surveillance that might become necessary in the future if there were to be great reliance on plutonium. If there were a significant number of factions or individuals who might be prepared to use plutonium in threats against society, then widespread surveillance could scarcely be avoided. We find it hard to believe that such an intolerable situation could arise in this country, though it might do so in countries with repressive régimes. It must be remembered, however, that in considering the hazards of the plutonium economy we are concerned with conditions as they might be fifty or more years ahead. What is most to be feared is an insidious growth in surveillance in response to a growing threat as the amount of plutonium in existence, and familiarity with its properties, increases; and the possibility that a single serious incident in the future might bring a realisation of the need to increase security measures and surveillance to a degree that would be regarded as wholly unacceptable, but which could not then be avoided because of the extent of our dependence on plutonium for energy supplies. The unquantifiable effects of the security measures that might become necessary in the plutonium economy of the future should be a major consideration in any decision concerning a substantial increase in the nuclear power programme.

333. The third area of concern is over the action that would need to be taken to recover plutonium that was known to have been stolen, or in the circumstances in which a nuclear threat had been made. There would be a wide-scale and determined search for the missing plutonium, and general search warrants (which are at present illegal) might be required. The degree of disturbance to people in the course of a nation-wide search, or in the evacuation of a threatened area, could be considerable. There might well also need to be restrictions on the rights of movement and assembly and the suspension of *habeas corpus* if the threat of the plutonium being exploded were serious. Such powers already exist in theory under the Emergency Powers Act, 1920, but there might be doubts whether the occasion justified their use. It should be noted that the problems of enforced evacuation would also arise if there were a serious release of radioactivity as a result of a reactor accident.

Security arrangements in the UK

334. As we stated at the start of this chapter, we have not attempted to assess the present security arrangements but rather to obtain an understanding of how the security problems that might arise through developments in nuclear power are viewed by the authorities concerned. We note here, however, that the visits we made to nuclear installations left us with the impression that insufficient attention had been given to some aspects of security. We were accordingly glad to learn from the discussions we have since had on security matters that the arrangements have been thoroughly reviewed and that improvements have been, and are being, introduced. One of these is that the UKAEA constabulary now have access to firearms and permanent statutory authorisation

for this has been provided through the Atomic Energy Authority (Special Constables) Act recently passed by Parliament. We accept the necessity for this measure though we share the disquiet that was expressed by some Members of Parliament during the debate on the Bill about the potential implications of this step.

335. We are confident that the security hazards associated with the present level of nuclear development in the UK are now fully appreciated by the Government and the authorities concerned, and that the security measures now in force or planned are adequate for present circumstances. We have no doubt that these measures will be periodically reviewed, and if necessary strengthened, in the light of nuclear and other developments that would affect assessment of the risks. However, a flexible response to security risks in the light of events is one thing; it is quite another to question whether the hazards of nuclear development in the future could become so great that adequate security could not be ensured, or alternatively whether the implications of the security measures needed could become unacceptable to society. We cannot see that the present system by which decisions are reached on nuclear development allows us to address ourselves to such questions.

336. The issues we have discussed in this Chapter—the risk of sabotage against nuclear installations, the risk of plutonium diversion and its use in terrorist action or threats against society, and the extent of the security measures that might become necessary to provide adequate safeguards—are by their nature very difficult to assess. The significance that they might assume in the future can be only a matter of opinion, depending on speculative judgements about likely developments in society, and to some degree in the world at large, which no one can make with certainty. Nevertheless, these issues are real and important and of a kind which, in our view, require wide appreciation and discussion. Public debate will not resolve them but it may form a climate of opinion which would assist Government in assessing the weight that should be given to these matters in decisions on nuclear development. Though serious risks from such development probably lie well into the future, judgment about their possible severity and acceptability could react on decisions that need to be taken now.

CHAPTER VIII

RADIOACTIVE WASTE MANAGEMENT

Introduction

337. We have described briefly in Chapter III how radioactive wastes are generated at various stages of the nuclear fuel cycle, and in Chapter V we discussed the administrative arrangements that exist to control and authorise the discharge of some categories of waste to the environment. In this chapter we consider the subject of radioactive wastes in more detail. We shall be mainly concerned with those wastes whose management presents particularly difficult problems and which are at present stored at nuclear sites, and with the steps that are being taken to ensure that they will not cause undue harm to the environment.

338. In Chapter IV we stated our view that there should be no commitment to a large programme of nuclear fission power until it has been demonstrated beyond reasonable doubt that a method exists to ensure the safe containment of long-lived, highly radioactive waste for the indefinite future. These wastes already exist in substantial quantity and a safe method for their long-term disposal is in any case required whatever is decided about nuclear development in the future. We are clear that such a demonstration will require a substantial programme of research. In the latter part of this chapter we consider the possibilities that exist for the safe disposal of these wastes and the organisation needed to pursue the the search and judge its results.

Gaseous waste

339. The principal radioactive discharges to atmosphere from nuclear reactors are of inert gases and of iodine. Iodine is a volatile fission product which escapes through leakage from some fuel elements: the isotope of principal concern is iodine–131 with a half-life of 8 days. There are two pathways back to man, direct inhalation and ingestion through milk, both of which give a radiation dose to the thyroid gland. However the routine discharges are small and the maximum annual thyroid dose is of the order of a few mrem, and usually far less. Iodine–131 would be one of the principal sources of danger to the public in the event of a large accidental release of radioactivity, and we described the probable effects in paragraphs 266–8. In the future, there may have to be controls on the emission of the very long-lived isotope iodine–129 (half-life, 16 million years) because of its persistence; this has already been proposed in the USA(35).

340. Inert gases disperse more widely into the atmosphere, and the collective population exposure is more important that the doses to a few individuals. The second largest component of the exposure of the public in the UK to radiation doses arising from waste disposals in connexion with the nuclear power programme comes from argon–41. In the six early Magnox stations with steel

pressure vessels, it results from neutron irradiation of argon, a minor constituent of the air used to cool them. Argon–41 is a β-emitter with a half-life of 110 minutes. Recent estimates of the effects of the discharges(7) are similar to the estimate previously given by the NRPB of a collective dose of 250 man-rem(4). It is an insignificant hazard to the public, and since all later reactors have concrete pressure vessels which are water-cooled and thus avoid the problem, it will not increase much in the future.

341. Two other isotopes of concern at reactor sites are tritium (hydrogen–3) and carbon–14, which is discharged in the form of carbon dioxide. (Each is also discharged at reprocessing plants, although to a lesser extent.) Reactors that use heavy water as a moderator form tritium by neutron capture in deuterium, and fast reactors are also expected to discharge the isotope. Special techniques will be needed for the monitoring of personnel in order to ensure that they are not excessively irradiated. Carbon–14 by contrast is produced in greater amounts by the gas-cooled reactors (Magnox and AGRs) because they employ carbon as a moderator which contains nitrogen as an impurity. The quantities of carbon–14 discharged are expected to be much less than those of tritium(36) but since its half-life is 5730 years it could present a long-term dose commitment. There is a large amount of naturally occurring carbon–14 in the atmosphere and biosphere, and a rapidly-expanding nuclear power programme might give rise to an increase of a few per cent by year 2000. This would be much smaller than the levels reached temporarily in the 1960s as a result of atmospheric bomb-tests.

342. At fuel reprocessing plants, the largest discharge of radioactivity to air is krypton–85, a fission product that escapes when the fuel is chopped up before being dissolved in acid. Even though more than 1 MCi is currently discharged each year from Windscale, the dispersion into the atmosphere is such that it gives doses to skin of the order of only about 1 mrem/year to those living near the plant, and these are considered acceptable. In the future, concern will attach to the collective dose received by people living in the region, or even to that of the world population because of the progressive build up of this inert gas in the atmosphere. The AEA have therefore anticipated this potential future problem, and have developed a means of removal and safe storage for the krypton–85 should this subsequently be needed (see paragraph 133). However, this is an international problem, and control measures will need to be implemented in all countries if they are to be effective.

343. It appears to us that the standards being applied to control gaseous emissions are such that there is no significant problem at present. Some potential areas of difficulty in the future are already being considered, and it appears that technical means of reducing discharges will be available when needed. The actual discharges are not published, and we have recommended in paragraph 251 that HMPI (or any other controlling inspectorate) should publish the discharges each year, with their estimated effects, in the same way as is currently done by MAFF for discharges to water.

344. In the future we expect that a rather more systematic approach to the control of emissions may be needed. In the White Paper "The Control of Radioactive Wastes" (Cmnd. 884, 1959), there are strict upper limits given to the doses that may be received by an individual or a population, and the principle is stated that doses should be reduced as far below these levels as is reasonably practicable having regard to cost, convenience and the national importance of this subject. This in effect says that the "best practicable means" (bpm) should be adopted to determine how far below the Derived Working Level (DWL) the discharges should be kept. (The DWL is that level of discharge that would, on average, just irradiate the most exposed members of the public to the permitted level, usually the one recommended by the ICRP.) We agree with this approach, but recommend that the agreed maximum level should then be regarded as a presumptive standard in the same way that the Alkali Inspectorate at present lays down such standards in non-nuclear industry, for example, for dust in emissions from a cement works. This standard would then form part of the overall authorisation to discharge, which at present is generally defined numerically only for aqueous effluents.

345. The Alkali Inspectorate have, in fact, adopted this approach in relation to the discharges to atmosphere from the AGRs, where the presumptive standard or authorised limit for discharges has been set at 5 per cent of the DWL([37]) for the two radioisotopes of principal concern. "Bpm" also embraces a number of other matters such as regular monitoring and maintenance. The authorised limits for any particular plant will need to take account of many factors, such as the ease with which the plant could reduce the discharge, and what would happen to the radioactivity not thereby discharged.

346. The authorised limits need to take account not only of the dose received by the critical group, who may be very few, but also of the collective population dose from the plant. They should reflect also the benefits conveyed by the plant, viz, its output of electricity. It is interesting that in the Scandinavian countries, this requirement is embodied in a formula; at present it is recommended that the collective whole body dose (exclusive of occupational exposure) from the full fuel cycle operations should not exceed 1 man-rem per MW of electrical capacity, of which no more than half shall be due to releases from nuclear power stations*. Finally, it will be necessary to consider the additive effects of radiation doses to individuals or to the population received from aqueous discharges from the plant, and also from discharges from other nuclear sites, some of which may be overseas†, in setting the authorised limits.

347. These requirements are multifarious and may, particularly in the future, prove difficult to balance. Nevertheless, we think that they should be examined in enough detail to allow each nuclear site to be given a clear standard to which

*Discharges in the UK would give about one-half of this recommended maximum level to people in the UK.

†For example, discharges of caesium from Windscale are reconcentrated in fish caught in the Irish Sea, and they give an estimated collective dose of 1900 man-rem to people in foreign countries who eat this fish.([6])

it should work. The only difference between limits for nuclear plant and for non-nuclear plant would be in the time interval over which they should be maintained. The discharges of radioactivity are presumed to have an equal effect if they are spread uniformly over a whole year or concentrated into a single day, and the monitoring system should therefore reflect the need to check the total activity emitted. But dust from a cement works, for example, could well be unendurable if all the emissions were concentrated into just a few days —particularly if they were Mondays.

Liquid waste

348. Nuclear power stations also give rise to radioactive aqueous effluents, but the quantities of radioactivity are normally quite small. In Magnox stations, the main source is from corrosion of the fuel element cladding in the water of the cooling pond (see paragraph 128), which is significant if there are delays in shipping out the irradiated fuel for reprocessing. Radioactive contamination of this kind has occurred, for example, at Hinkley Point, and caused considerable difficulties for the operators. The pond water can be treated to reduce the discharges, but this process creates a wet radioactive sludge which in turn requires some form of storage and eventual disposal. Thus the price of a high standard of aqueous effluent from a power sation may well be a sizeable on-site storage problem for solid or semi-liquid waste. We return to this issue below.

349. Another liquid radioactive waste formed at nuclear power stations is used lubricating oil. The CEGB have arranged to burn this as a small proportion of the feed at some of their oil-burning power stations, which is an eminently sensible arrangement as the amount of radioactivity is quite small and it is thereby well dispersed. We mention this only to show the pervasive nature of radioactivity, and the consequent need to devise suitable management techniques for articles or materials contaminated with it.

350. By far the largest aqueous discharge of radioactivity to British waters takes place from BNFL's Windscale works, where over three-quarters of the total for the UK is authorised. The liquid discharged to sea, through twin pipelines some $2\frac{1}{2}$ km long, consists of low activity effluents, some originally only slightly radioactive, and some resulting from medium-activity arisings which have been stored in tanks for several years to allow their activity (much of which is short-lived) to decay to low levels. The low activity waste includes the effluent from the works laundry and the runoff of rainwater from the whole site. These are mixed with water from the irradiated fuel storage ponds and various process liquors such as evaporator distillates before discharge through the pipelines. The limits on discharge of liquid waste are quite complex, with an overall limit of 75 kCi of β-emitters in any quarter and lower limits on particular components such as ruthenium–106, strontium–90 and cerium–144 which are biologically reconcentrated in foodstuffs. There are inevitably some actinides such as plutonium in the effluent, and there is a limit on α-emitters of 6 kCi in any year (and 2 kCi in any four months), which was raised from 1800 Ci per year in 1970.

351. The discharge limits are currently being revised to take account of the changing importance of different environmental pathways, the technical possibilities for improving effluent standards, and the needs of BNFL over the next 10–15 years with an increased volume of business. Environmental pathways can change for a number of reasons. We explained in paragraph 80 that the concentration of ruthenium–106 in edible seaweed was no longer of prime concern, and that the most critical isotope is now caesium–137 which is concentrated into fish, especially plaice, dab and skate. A revised authorisation is in course of preparation, and will for the first time specifically limit the discharge of this isotope, and of the shorter-lived caesium–134. Such changes reflect careful monitoring activity and continuing research. What is more difficult to predict is the future movements of very long-lived radioisotopes, in particular plutonium and americium.

352. Nearly all the plutonium that is currently discharged to sea at Windscale ends up on the bottom sediments([38]). Some of these move into the Ravenglass estuary. They are thought to be radiologically insignificant at present, but it is known that the estuary contours are changing with time because of the net landward movement of sediment. Thus in about a century our descendants may be faced with two new exposure pathways for this plutonium. Sediments that are above high tide level and are no longer wetted may blow about in dry, windy, weather and possibly form respirable aerosols. A smaller hazard could arise from grass growing on newly reclaimed land, which may support cattle as happens now. They would require careful monitoring to ensure that there is no radiological hazard.

353. It has recently been suggested([39])([40]) that the plutonium in these sediments could be mobilised rather more quickly as a result of chemical reactions within the sediments and the resuspension and transport ashore of fine particles as a result of storms. Chemical changes are produced by the progressive depletion with depth of the oxygen content of the water within the sediment. This may cause the isotopes to become much more chemically mobile. In some areas the activities of burrowing creatures can bring heavy metals to the surface of the sediment even if they were once securely buried. Storms and tides might then resuspend the fine particles from the upper surface of the sediment. Once in suspension, the particles might be carried considerable distances and storm conditions could assist in bringing them ashore rapidly. Not enough is yet known about such processes and more needs to be learned about the conditions under which a hazard might arise from the plutonium and americium now in the Irish Sea sediments. It will be necessary to complement further field observations with a programme of laboratory work.

354. We have concluded that there is insufficient evidence at present to require the discharges of plutonium from Windscale to be reduced. Nevertheless this is a serious issue for the future and since two possible environmental pathways have already been recognised we consider that the matter should receive more attention. A wide range of options needs to be explored, such as the consequences for BNFL if it were found necessary to impose much stricter limits on discharges. Since the processes occurring in the bottom sediments are of

principal concern, the scientific work will need to call upon the expertise in sedimentology that exists within some universities and NERC institutions. Studies will need to be undertaken of the mechanism by which individual environmental factors influence the adsorption of plutonium onto and its desorption from sediment particles. These factors include the presence or absence of organic material and fine-grained iron and manganese oxides, and the ion exchange capacity of the sediment particles of various sizes under different conditions of oxygenation and temperature. Such laboratory work will also be highly relevant to the programme needed to evaluate the safety of disposing of high-level wastes by burial within the ocean bed which we discuss further below (paragraphs 409–418). Research should also be conducted on the possible countermeasures that could be needed if the plutonium-bearing sediments were to migrate landwards and move upwards to the tidal range or above it. These might include a programme of dredging, with the spoil being deposited in the deep ocean. The problem is essentially multi-disciplinary and besides MAFF will need to involve the NERC and the NRPB, together with experts in the aerial resuspension of fine dry particles.

Solid waste (low level)

355. Solid radioactive wastes are generated at all stages of the nuclear fuel cycle, but before discussing them, we briefly mention the arrangements made for the control of radioactivity in very small amounts which arises from activities other than nuclear power. Solid waste which is contaminated with small amounts of radioactivity is permitted by the DOE to be deposited on municipal and other refuse tips. There are two classes—ordinary tips and tips where "special precautions" are applied. All refuse tips are considered to be suitable for the disposal of very small (microcurie) amounts of radioactive waste admixed with general refuse, such as would arise from luminous articles and electronic valves. "Special precautions" tips owned by county councils exist in some 18 counties in England, and there are eleven privately-owned tips. There are two tips in Wales. They are permitted to accept only sealed containers, each of which may contain one millicurie of short-lived (half-life less than a year) radioisotopes or one-tenth this amount of longer-lived ones such as carbon–14 and tritium.

356. The disposals of radioactive material on "special precautions" sites are small, and care is taken to ensure that the geology is suitable and the tip is well managed. They are radiologically quite innocuous but nevertheless there is sometimes strenuous local opposition. We have seen a number of press reports during the course of our study that describe concern about these tips. The subject is peripheral to the main issues of our study, but we thought it worthwhile to make some inquiries. As a result we have concluded that the present practices of the DOE's Radio-Chemical Inspectorate (RCI) in this respect are perfectly satisfactory. Moves to have slightly radioactive material carted across the country at great expense and then dumped in the ocean, after the proposals for local dumping have been approved by the local and regional water authorities, strike us as misguided.

357. The lowest activity wastes arising in this country from the nuclear fuel cycle comes from uranium conversion and fabrication works, Springfields and

Capenhurst, and they are of concern more because of the chemical toxicity of uranium and its compounds than through their radioactivity. Nevertheless, they require disposal and BNFL maintain a site at Ulnes Walton, Lancashire, for the purpose. Since these disposals are from licensed nuclear sites, they require the joint approval of DOE and MAFF (in Scotland the SDD alone, advised by the FRL). The same control arrangements apply to the site at Drigg near Windscale, a stretch of scrubland where active waste is buried in trenches. This site is used by BNFL mainly for their own low-active waste—typically contaminated laboratory equipment, protective clothing and miscellaneous rubbish—but they also accept small amounts from outside organisations. The authorisations are given in terms of activity concentrations and surface dose rates rather than total amounts of activity but the actual amount buried each year at Drigg is quite small, about 100 curies of total activity in a volume of some 8,000 cubic metres.

358. The Drigg site also contains a number of concrete bunkers originally built during the war for the storage of explosives. Some of them are used as temporary stores for drums of contaminated waste that arose from the nuclear weapons programme in the 1950s. These have started to corrode, and it was announced in April that they would be removed from Drigg for disposal by ocean dumping. However, the bunkers are also being used for the storage of plutonium-contaminated solid waste, much of it arising from the plutonium fuel element factory at Windscale. This, too, is ultimately destined for disposal in the Atlantic. At present there is rather more than 3,000 cubic metres of such material, containing about a third of a tonne of plutonium. Future arisings are expected to be on a smaller scale because of progress in the development of waste management techniques, notably the incineration of α–active material, and the subsequent extraction of plutonium by leaching it out from the ash. The new "wet" process of making mixed oxide fuel pellets (see paragraph 144) will also reduce the amount of plutonium-bearing waste needing to be stored.

359. We visited the Drigg site in the course of our visit to Windscale. The burial of waste is inevitably a somewhat untidy operation and the arrangements at Drigg were in marked contrast to those at a disposal site we visited at Asse in West Germany. Here the waste is buried in a disused salt and potash mine. At present both low-level wastes (in drums requiring no additional shielding) and intermediate-level wastes are accepted and stored in large previously-excavated caverns which can then be backfilled with salt. The work takes place between 500 and 700 metres underground and incidentally in very comfortable conditions for the operators. There is confidence that the salt, whose existence in a massive formation is evidence that groundwater has not been circulating for several million years, will continue to isolate the wastes from the biosphere for a comparable period in the future. The only geological disposal site in Britain is a shaft on the AEA site at Dounreay, which was authorised by the SDD for the disposal of a number of active items. The Institute of Geological Sciences (IGS) were commissioned by the AEA in 1967 to examine the possibility of extending the facilities.

360. We are assured that the waste buried at Drigg poses no danger to the public. The RCI are confident that no leachate from the trenches reaches the underlying aquifer; surface run-off goes to a local stream whose water continues to be safe to drink. But we are doubtful whether a relatively simple operation such as this should be replicated in a series of sites around the country as might be needed to deal with the waste from a very big nuclear power programme in the future or if it were decided that the solid waste now accumulating at nuclear power stations should be buried. We consider that more thought needs to be given to the arrangements that may be needed in the future. Geological conditions and available disposal sites are clearly different in Britain from those in West Germany, and it would not necessarily be appropriate to try to duplicate the Asse facility. We refer later (paragraph 367), however, to the need to develop a national disposal facility for other kinds of wastes that would offer comparable security to that of the Asse mine. Such a facility should also be able to handle arisings of low level wastes in the future.

Solid waste (intermediate level)

361. For the purposes of discussion, we include in this category all solid radioactive waste other than that which can be accepted for land burial and other than the high level fission products, which will eventually be solidified. These wastes do not require special cooling to remove fission product decay heat, but many of them contain high energy γ–emitters and therefore cannot just be drummed without additional external shielding to protect the operators. Wastes of all the types considered in this section can in principle be accepted and buried at the Asse mine.

362. *CEGB and SSEB sites.* At each of the Magnox and AGR sites, the CEGB expect that some 2,400 cubic metres of radioactive solid waste will be accumulated over the 25-year life of the stations. At the end of this time the activity would amount to about 80 kCi, decaying to half this level a decade later. At present the CEGB are storing about 7,000 cubic metres in total, which is a volume equivalent to that of three average semi-detached houses on each nuclear site. Half the total volume, containing the major part of the radioactivity, comprises metallic components that have been removed from the reactor core such as fuel stringer materials, control rods, chains and neutron flux measuring instruments. These are kept in concrete storage vaults within the reactor's biological shield as they are intensely radioactive, mainly because of cobalt–60 from neutron irradiation of steel. Cobalt–60 emits penetrating γ–rays, but as its half-life is only 5·3 years, it will effectively decay over periods of several decades—but not before the reactor site may need to be cleared for a new type of reactor. This possibility in fact plays (or should play) an important role in determining power station design as well as waste management strategy. The UK, being small and densely populated, has relatively few sites that meet the stringent requirements for nuclear power stations, and it is likely that existing sites will need to be reused. This would be easier if there were no large accumulations of radioactive waste at the stations.

363. The other wastes from Magnox and AGR stations will consist mainly of miscellaneous incombustible materials such as filters, valves, ash from the

incineration of combustible waste, and dust. Some combustible items may be unsuited to incineration, and will have to be treated as solid waste. There is a relatively small volume of wet material, such as sludges from spent fuel cooling ponds and ion-exchange resins used to treat pond water before discharge.

364. We have been unable to discover any clearly formulated policy for the future disposal of this waste. The utilities are permitted to keep it on their sites by the NII, except for a limited quantity of material from cooling ponds which is contaminated with plutonium, and which is to be disposed of within five years (although we know of no plans to do this). The waste is quite secure at present, but presumably it will have to be removed eventually if the reactor sites are to be re-used. CEGB do not wish to operate a disposal facility themselves as they consider it inappropriate. BNFL are reluctant to accept the waste at Drigg, and anyway most of it would be outside the authorised limits for that site. Some of it could be packaged and dumped at sea, and indeed the NRPB told us in evidence that they thought this would be both appropriate and safe. Sea disposal might be expensive, but in any event we are not aware that the possibility is being actively pursued by the utilities. The impression we have formed is that there is a lack of clarity about where responsibility rests for determining the best strategy for dealing with these (and other) wastes and for specifying the practices that should be followed.

365. *AEA and BNFL sites.* The distinguishing feature of active waste on these sites is that most of it is contaminated with plutonium. Whereas much of the power station waste will decay to harmless levels in periods of the order of a century, waste contaminated with plutonium (except in trivial amounts) must be permanently removed from man's environment. The problem of disposal is therefore comparable with that of the high level fission products, except that there is generally little problem of removing decay heat because the activity is contained in much greater volumes. The total quantitites being stored by BNFL now, and estimated by them for 1985, are shown in Table 9 below, with the corresponding activities in high-level liquid wastes.

TABLE 9

Volumes and activities of various types of solid waste accumulated at Windscale in 1974 and estimated for 1985. The comparable figures for high-level liquid waste concentrate are also given

Type of Waste	*Volume, m^3*		*α–activity, kCi*		*β–activity, MCi*	
	1974	*1985*	*1974*	*1985*	*1974*	*1985*
Fuel element cladding . .	4,000	13,000	10	100	1	15
Sludge from treatment of liquid waste (obsolete process)	5,000	5,000	10	10	—	—
Plutonium-contaminated (low βγ content)	3,000	5,000	60	100	—	—
High-level liquids	550	1,800	500	2,000	350	4,500

Thus the amount of solid waste, all of it significantly contaminated with plutonium, being stored currently at Windscale is some 12,000 cubic metres, and within this volume there is a little under half a tonne of plutonium. There is also a growing amount of americium and in the future, as fuels of higher burn-up are reprocessed, the α–activity will be increasingly from this element.

366. The most rapidly growing fraction of this waste is the fuel element cladding. This is the outer covering of the fuel pins which is stripped or chopped before the fuel is dissolved in hot nitric acid, as described in paragraph 133. Most of the present cladding material is magnox, which is potentially inflammable, so it is stored under water in concrete silos. Zircaloy hulls from LWR fuel are also stored here*; they tend to contain considerable quantitites of tritium, a tertiary fission product, which is released to the water. The silos are used to store miscellaneous active hardware and the insoluble residues from high burn-up oxide fuels. These have a very high specific activity and may increase significantly in fast reactor fuels.

367. It may prove difficult in the future to retrieve and package this material for ultimate disposal. In particular, the presence of considerable quantities of water with high radioactivity content represents a hazard. The AEA are investigating the possibility of decontaminating the fuel cladding hulls so that they would be less radioactive. This would facilitate disposal although the radioactivity removed would remain to be disposed of elsewhere, probably with the high-level waste. The present policy is to allow the waste to accumulate so that the levels of radioactivity may decline before it is drummed for ocean disposal, starting in 1990. For the reasons given below (paragraph 372) we think that great difficulties may be encountered in implementing this policy. We see a need for a national disposal facility able to accept these wastes, and the operation of such a facility would be one of the functions of the new organisation we propose in paragraph 432. New arisings of waste would be suitably encapsulated so that the radioactivity is effectively immobilised, and transported as soon as possible to the disposal site. This appears to be the policy in West Germany, and all the solid waste taken for disposal at Asse is embedded in concrete or bitumen within the steel drums.

368. The Windscale plant deals routinely with highly radioactive and dangerous materials and it is not surprising that incidents that inevitably occur should receive much, and sometimes excessive, attention. Small leaks of radioactive material whose radiological implications are trivial have received exaggerated attention in the press. We have no reason to think that the operations at Windscale are not conducted with every attention to safety. Nevertheless, it is important at such a plant that the highest standards of general housekeeping should be employed and we feel bound to say that we did not gain the impression that this was so at the time of our visit (November 1974). We would urge that this aspect should be given more attention by the new management at the plant.

369. *Ocean dumping.* It is the policy of BNFL (as expressed in their evidence to us) that solid wastes should be disposed of by dumping in the deep ocean,

*Stainless steel hulls (from AGR fuel elements) are kept in dry silos.

other than those that are sufficiently innocuous to be buried at Ulnes Walton and Drigg. However, no radioactive wastes can be disposed of from nuclear installations within the UK without the approval of MAFF. The London Convention, which entered into force in September 1975, restricts the sovereignty of states with regard to ocean dumping and prohibits the dumping of high level wastes. The Convention requires that there should be no dumping without a prior authorisation from competent national authorities, and that before giving such authorisation these authorities should give careful consideration to the environmental effects. The Convention encourages collective action through appropriate international bodies to that end. The UK takes part in an annual international operation to dump packaged, low-level solid waste in the deep North Atlantic (the dumping area is about 900 km SSW off Land's End in some 4,500 metres of water). Since 1967 these operations have been conducted and controlled by the Nuclear Energy Agency (NEA) of the OECD on behalf of member states who wish to dump, an arrangement that is politically and technically convenient to the states concerned.

370. Over many years the IAEA have established criteria and standards as the bases for recommendations made by expert panels on various aspects of waste disposal, including ocean dumping. The NEA have also carried out special studies with regard to their own operations. Following the London Convention, the IAEA were asked to define the high-level wastes whose dumping at sea would be prohibited. The IAEA definition is expressed in terms of concentrations in curies per unit gross mass for particular radioisotopes; the definition is based on an assumed upper limit of 100,000 tonnes per year at any one site. The definition is an interim one, subject to review every two years in the light of new knowledge. In arriving at the definition the IAEA used the oceangraphic model for the North Atlantic adopted by the NEA and substantially prepared by Webb and Morley of the NRPB[41]. The model deals with the processes of dilution and dispersion of radioactivity in the ocean and the likely reconcentration factors in fish and other marine organisms, and is used to calculate the rate of dumping which would eventually cause irradiation of members of the public at the ICRP dose limit. The resulting limits of activity per tonne dumped (most of which will be concrete in which the wastes are embedded) are as shown in the following Table:—

TABLE 10

Levels of activity permitted by the IAEA in solid waste dumped in the ocean (under review)

Radioisotope	*IAEA recommendation*		*Webb and Morley calculation*	
	Ci/Tonne	*Ci/year*	*Ci/year*	*Factor of Safety*
α–emitters radium	10	10^2	10^6	10^4
plutonium	10	10^6	10^{10}	10^4
Strontium–90, Caesium–137	10^2	10^7	10^{11}	10^4
Other β–emitters	10^3	10^8	10^{12}	10^4
Tritium	10^6	10^{11}	10^{15}	10^4

371. The factors of safety given in the last column are those by which the permitted release of the particular radioisotope would have to be increased to irradiate someone to the level recommended by the ICRP as the maximum that should be accepted. These would theoretically apply if the full 100,000 tonnes were dumped each year in one location and if it was all at the IAEA limit on specific activity. Webb and Morley list a number of additional factors which combine to increase the safety margin by a factor of 10,000 (100 for tritium). One factor allowed for in the analysis is the possibility that diffusion and weak currents will transport masses of water containing the radioactivity towards fishing grounds where the activity might be concentrated rapidly in the food chain. We have been advised by the Institute of Oceanographic Sciences (IOS) that this part of the analysis is suspect as little is known about such processes and they may well be of great importance. Moreover, it has since been shown(42) that the reconcentration factor for plutonium in fish and molluscs is about an order of magnitude higher than the figure appropriate to plutonium originating from bomb-test fallout, which was used by Webb and Morley. In the future, americium will be present in increasing amounts in solid wastes, and it is known to be significantly more biologically mobile than plutonium(43).

372. We cannot, therefore, regard the present analysis which defines the limits for ocean dumping as offering a secure basis for large dumping operations in the future; no doubt the IAEA will have these and other relevant factors in mind in keeping the limits under review. We understand that MAFF, the IOS and NRPB are at present developing new models. It does not seem that the present scale of dumping in the North Atlantic, amounting to about 7,000 tonnes per year, poses any worry. The amount to be dumped is determined by the governing body of the NEA on the basis of individual countries' needs and what is considered to be politically acceptable. At present, the main participating countries are the UK, Belgium, the Netherlands and Switzerland, and up to the present it has been possible to accommodate their requirements. It is by no means clear, however, that this will be so in the future because of the increasing concern of some member states about the principle of ocean dumping. But BNFL hope to be given permission to dump some 2MCi/year of β–emitters (most of which would be the relatively long-lived strontium–90 and caesium–137) and 24kCi/year of α–activity in the 1990s. These activities are each about two orders of magnitude greater than those currently being dumped by the UK. We think there are very strong grounds for doubting whether dumping on this scale will be acceptable to the UK authorities or internationally, and it is for this reason in particular that we have foreseen the need (paragraph 367) to develop a national disposal facility.

High level liquid waste: interim management

373. The rest of this chapter is devoted to the problems of management of the high level wastes, mainly fission products, which are at present being stored at Windscale (with smaller quantities at Dounreay) in special tanks which can remove the internal heat from their radioactive decay and keep the solids that precipitate out in suspension. We gave a brief account of these wastes

in paragraphs 137–141, and Figures 9 and 10 show the quantities of radioactivity involved. In this chapter we shall describe a number of technical options for their management, and discuss their implications for research work, and for the development of nuclear power. These wastes are the most intractable of all those produced in the nuclear fuel cycle, and contain the overwhelming preponderance of the activity (except that of plutonium). If satisfactory means for their disposal can be found, then in principle the same method would be adequate for the other wastes that we have discussed so far.

374. Our discussion of these wastes is in two parts. The first deals with broad questions of whether the wastes should be produced in this form at all, and with whether it might be possible to dilute them in the oceans so that they did not need to be stored at such expense. We then go on to describe the process at present proposed for their vitrification, that is, for their incorporation in glass, and discuss its constraints. Finally we examine a particular problem, that posed by the actinides, mainly plutonium, americium and curium, and a possible approach to its solution. In the second part, we discuss the possibilities for long-term storage and disposal of the waste (which are assumed to have been solidified) and what needs to be done to provide assurance that a satisfactory means of disposal has been developed.

375. *Waste formation.* Any modern discussion of waste management normally begins with consideration of how the generation of the wastes might be avoided, or minimised. The quantity of fission products produced is roughly proportional to the electricity generated in nuclear power stations, and does not greatly depend on the type of reactor or its operation. The fission products, once formed, decay, with their characteristic rates. Although in theory some of them could be irradiated in a reactor to give isotopes with shorter half-lives, this is not feasible in practice. (However the higher actinides, americium and curium, could possibly be treated in this way, and we discuss this idea below). The only form of nuclear power that would generate very much less residual radioactivity would be thermonuclear fusion (see paragraph 152). Its advantage from the point of view of radioactive waste management is a strong argument for supporting its development.

376. Since many of the most troublesome problems of radioactive waste management arise as a result of the reprocessing operation, it is natural to consider whether it would be preferable to leave the spent fuel unreprocessed, to store it until its heat generation rate was quite low, and then to dispose of it to a suitable repository. This possibility has been advocated by some who would thereby plan to avoid the problems of separated plutonium and the need for safeguards. The complete abandonment of reprocessing would imply the abandonment of fast reactors which require the plutonium as fuel; severe limits would then be set on nuclear expansion given the probable extent of uranium reserves (unless the large quantities of uranium present in the oceans could be extracted). A more prudent approach might be to provide for the storage of irradiated fuel in such a way that it could be retrieved for later re-processing if this became necessary.

377. The AEA gave evidence to us on the implications of a decision to delay or abandon reprocessing. They commented that nuclear fuel elements were designed to give a good performance in the particular environment of a reactor and that it was not to be expected that they would retain their integrity over very long periods of time in storage. Some fuel elements would inevitably be leaky and require separate treatment. Fuel clad in stainless steel or zircaloy could be stored for a few decades in ponds but Magnox fuel corrodes rapidly in water. In the longer term some form of "bottling" or encapsulation of the fuel elements would be needed and this could be just as difficult as reprocessing, although routine releases to the environment would be less. Several techniques appear to be feasible which could provide for the long-term storage of fuel elements and for their eventual retrieval for reprocessing. Later in this chapter (see paragraph 389) we discuss ideas for the ultimate disposal of high-level radioactive wastes with the object of isolating them permanently from man's environment and making them virtually irretrievable. In order to achieve these objectives some degree of reprocessing would be essential. It is possible, however, that one of the techniques that is being investigated for ultimate disposal, that is, disposal to geological formations on land, would also be relevant to the construction of a highly secure and very long-term storage facility.

378. The question of whether reprocessing could eventually be delayed or abandoned depends on many factors which cannot be fully evaluated at present. Major factors are the extent of the future need to deploy fast reactors, given that these are successfully developed, and the outcome of research programmes into techniques for the ultimate disposal of high-level wastes. There appears to be no real alternative to the reprocessing of Magnox fuels, and the techniques for reprocessing oxide fuels must certainly be developed against future needs. But if it were eventually decided that the use of fast reactors could and should be avoided or indefinitely postponed, and if it were found that geological formations offered the high degree of security that would be required, the option of not reprocessing the fuel elements is one that should then be seriously evaluated.

379. The design of fuel reprocessing plant depends quite significantly upon the level of radioactivity in the fuel, and hence upon the time delay between discharge from the reactor and commencement of reprocessing. For Magnox fuel this is about 6–12 months, and a major constraint has been to avoid corrosion of the cladding which becomes very awkward to deal with after a delay of more than two years. With oxide fuel in stainless steel or zircaloy cladding, this problem does not arise, and indeed large quantities of irradiated LWR fuel are being stored in ponds for many years in other countries because of the lack of any commercial reprocessing plant. In the future, if fast reactors become the dominant type, there will once again be a need to reprocess quickly because of the value of the plutonium, and it might pay to allow a cooling period of much less than the 6 months which is assumed adequate for the iodine–131 to decay to negligible levels*. On the other hand, a delay of several years

*For example, the USAEC draft EIS on the LMFBR (WASH–1535) based its economic analysis on a 120-day cooling period.

before reprocessing would allow most of the curium formed in the reactor to decay to plutonium and thus be removed from the wastes. Thus environmental considerations play, or should play, an important part in the determination of reprocessing strategy.

380. We referred in paragraph 139 to the need to provide secure containment for the high level wastes because of the possible consequences of even a small leak from the Windscale site into the Irish Sea. Discharge into the main body of the Atlantic Ocean might be thought to offer a sufficient degree of dilution, although, as we have seen, this is prohibited under the London Convention. It is of interest to note that it has been calculated([44]) that a world fission power programme rising to 1,200 GW in year 2000 would by then have generated such large quantities of fission products (and actinides) that even if they were dispersed uniformly in the vast bulk of the oceans, the resulting concentration would be within one or two orders of magnitude of the maximum permissible concentration for drinking water. This would not be satisfactory because of the many food chains that could concentrate the radioisotopes and return them to man.

381. *Vitrification of wastes.* We described the processes by which the concentrated liquid waste was formed in paragraph 140. The constraints on the amount that can be stored in one of the tanks are the heat removal capacity and the quantity of solids. The latter remain approximately constant, but the heat generated by a given quantity of fuel decreases with time as shown in Figure 8*. There is a sharp decrease between 6 months and 1 year, and thereafter a steady decline over the next few years. Accordingly, parallel filling of tanks is practised and the average heat rating of the liquid in the tanks is much less than that of the latest addition. BNFL plan to vitrify the wastes using a process whose development started at Harwell in the late 1950s, but it will not be in commercial operation until 1985. The delay in bringing the vitrification process into commercial production stems from a long period of inactivity in the 1960s when no further development work was carried out. It is strange in retrospect that a matter so important for the safe development of nuclear power should have been delayed for so long. On the assumption that the commercial-scale plant is in operation by 1985, BNFL plan to work through the backlog by 1995, but there will always be a requirement for a few tanks in order to provide a buffer capacity between the reprocessing and vitrification plants.

382. The vitrification process consists in running a measured amount of high-level liquid waste concentrate into a stainless steel crucible where it is heated together with glass-making materials, silica and borax. When the glass has formed, the filling ports are welded over and the crucible becomes a permanent container. The fission products may account for up to one quarter of the weight of the vitrified mass provided that they are not so "young" that their heat flux would melt the glass despite the forced cooling in the store. Thus a decision to reprocess the fuel early, perhaps because of the high value of the associated plutonium, might involve the manufacture (and subsequently, the

*The curves include the heating from the plutonium removed during reprocessing but this is relatively small over the period considered.

need for disposal) of more glass blocks than if this were delayed for a few years. Since vitrification is intended not only to facilitate ultimate disposal of the wastes, but also to make them more secure than the concentrated high-level liquids in the interim, it is desirable for the process to be able to accept wastes direct from the reprocessing plant without intermediate storage as liquid. Even a two-year delay before vitrification would largely nullify the short-term advantages of the process as two-thirds of the total cumulative radioactivity from a large reprocessing programme would then be contained in the potentially vulnerable liquid form([45]).

383. Another decision to be made is on the size and shape of the glass blocks. The original FINGAL process at Harwell produced cylinders 300 mm diameter and 3 mm long. The French, who have also developed the vitrification process, are building a plant at Marcoule which will be in service next year (1977) and will make short stubby cylinders, 500 mm diameter and 750 mm long.* A second, similar, plant will be built at their reprocessing works at Cap de la Hague on the Channel coast in 1978. BNFL, however, favour large annular blocks which would be better able to dissipate heat in the water-filled ponds that would be used to cool them prior to ultimate disposal, and could be produced in a single production line rather than a number of separate ones. They envisage that their plant, using the process now called HARVEST, would produce large annuli 1·3 m in outside diameter and nearly 3 m high with a mass of perhaps 15 tonnes. Each would contain the wastes from reactors that had generated some 10 TWh† of electricity: the present Magnox reactors would therefore require only about three such annuli per year to accommodate their wastes. But such large blocks would be awkward to transport to any final disposal site, and if disposal involved their being placed in drilled holes, these would not only have to be of large diameter and therefore very expensive, but also have provision for filling up the central hole with a suitable material. In practice it would probably be found convenient to remelt the glass and cast it into a more suitable shape for final disposal. This provides another example of the importance of waste management strategy for the decisions taken on details of the processes involved.

384. *Actinide separation.* The high-level wastes contain, as we mentioned in paragraph 137, both fission products and actinides. The principal fission products of concern after about seven years' cooling are strontium–90 and caesium–137, and since they each have half-lives of about 30 years, they remain important for several centuries. But the hazard from the fission products drops in less than a thousand years to below that of the ore containing the uranium used as fuel, when allowance is made for the activity of its associated daughter products. They could therefore, in theory, then be buried in a uranium mine. Plutonium and americium, however, have isotopes with half-lives measured in thousands of years and if they were left in the wastes in significant amounts would make them hazardous for tens or hundreds of thousands of years. Beyond this time

---

*These cylinders will contain wastes cooled for many years. "Young" wastes (less than one year out of the reactor) would require cylinders of much smaller diameter.

† 1 TWh=10^9 kWh (see paragraph 438).

the main hazard is from neptunium and its decay daughters. It is conceivable that engineered storage could be provided that would contain the fission products effectively for periods of up to 1,000 years (some buildings in current use are that old), but it could not outlive the actinides. The problems of ensuring the safe containment of high-level waste would be greatly eased if the actinides could be removed.

385. It has therefore been proposed(46) that the high-level liquid waste should be separated into two streams, one containing just the fission products which could then be vitrified and the other the actinides, notably americium. The actinides would be incorporated in fuel elements and fissioned in a fast reactor to give the normal range of fission products; the process is referred to as actinide incineration. Thus the problem of containment of the wastes would be reduced to that of keeping them secure for hundreds of years rather than for hundreds of thousands.

386. There are difficulties in making very accurate calculations on how the actinide fuel elements would behave in the reactor, but the AEA have informed us that the elements would need to be subjected to irradiation for about 30 years in order to convert 99·9 per cent of the recycled actinides into fission products. The special actinide fuel elements would require to be removed periodically from the reactor during this time for reprocessing, and this operation, as well as that of fabricating the fuel elements, would pose additional hazards of occupational exposure to radiation.

387. Moreover, there are very severe technical problems in separating the fission products and actinides to the degree that would be needed, and the process might itself generate substantial quantities of intermediate-level wastes. Americium and curium have similar chemical properties to the lanthanides, or rare earth elements, which make up about one half of the total mass of fission products. The separation would be harder in the early years as the lanthanides then contain about one third of the total fission product activity, whereas this fraction would be far less if the wastes were first allowed to decay for several decades. It might, therefore, possibly be worthwhile to separate the lanthanides plus actinides first so that the remainder of the fission products could be solidified; the actinides could then be separated from the lanthanides later if desired. Altogether the AEA estimated that the successful development of both the chemical separation process and plant and of the fuel and fuel fabrication process would take 15–20 years, and would cost tens of millions of pounds or more. The AEA are continuing their study of the potential of actinide incineration in order to obtain the data that are needed to assess whether these developments would be justified. Work is also in progress in other countries and the EEC has asked the AEA to undertake a study as part of a European programme on the handling and storage of radioactive wastes.

388. It is clear that the benefits that may be obtained from actinide separation and incineration are very uncertain; they will depend, for one thing, on the

outcome of research into the possibilities for permanent disposal of high-level wastes. We have formed the view that it would probably not be right to delay the programme of vitrification in the hope that the process might eventually be developed. We emphasise, however, that this is a tentative view and that because of the highly technical issues involved we should not regard ourselves as competent to make a firm recommendation. Decisions of this kind, which may have wide repercussions for the overall development of the nuclear fuel cycle, must be made by an expert body. The choice of the storage or disposal option to be used must take account of the technical features of the waste, and even the early stages of reprocessing may be dependent upon the ultimate means of disposal. We stress again the need for a clear strategy for waste management, and the importance of interaction between those whose concern is to protect the environment from the wastes and those who decide upon their management. We cannot see that, under the present arrangements, the many factors involved can be effectively brought together to formulate a definite strategy, and we return to this point in paragraphs 428–9.

High level waste: ultimate disposal

389. *The range of options.* The following paragraphs are addressed to the problem of isolating the high-level wastes, which are assumed to have been solidified into blocks of glass sheathed in stainless steel. We regard it as totally unacceptable that high-level wastes still in liquid form should be disposed of to a geological formation. The practice would be especially dangerous in the UK because of the relatively small scale of our geological formations and the occurrence of structural disturbances. We have assumed that the actinides will not have been removed, and therefore that it is necessary to look further ahead than the few hundred years needed to allow the majority of the fission products to decay to harmless levels. The wastes will generate a substantial amount of heat* which must be dissipated if the blocks are not to overheat and allow the glass to melt, and this must be considered in the choice of a geological formation.

390. We shall discuss a number of possibilities that have been proposed for the permanent and irretrievable disposal of the vitrified blocks containing the fission product wastes. To some, the concept of irretrievable disposal as compared with indefinite storage may smack of irresponsibility because it implies loss of control. On the other hand, any storage system implies some degree of maintenance and could be regarded as an imposition on our descendants who would have received no direct benefit from the power generated in reactors that produced the wastes. In addition, the storage of such immense quantities of radioactivity presents dangers from accidents or from malefactors or act of war. Engineered surface storage facilities might also be vulnerable to major climatic changes. They may serve adequately for a limited time, but they may also by their mere existence divert attention from the need for a permanent and final

*The heat from blocks containing 25 per cent fission product waste solids falls from about 60 watts/litre at one year to 10 watts/litre at seven years and 1 watt/litre at a century after the fuel has been discharged.

resting place for the wastes. Just as the existing system, whereby the wastes are stored in high-integrity stainless steel tanks with elaborate arrangements for cooling, is satisfactory enough at the moment but recognised as in no sense a final solution, so a system of engineered storage would face a similar charge.

391. Neither the AEA nor BNFL in their submissions to us gave any indication that they regarded the search for a means of final disposal of highly active waste as at all pressing, and it appears that they have only recently taken firm steps towards seeking solutions. We think that quite inadequate attention has been given to this matter, and we find this the more surprising in view of the large nuclear programmes that both bodies envisage for the coming decades, which would give rise to much greater quantities of waste. In the last year, and possibly prompted by our own inquiry or by the availability of funds for the purpose from the European Commission in Brussels, the AEA have commissioned the Institute of Geological Sciences (IGS) to conduct a desk study of the possibilities of disposal to a geological formation on land([47]). BNFL have asked the NRPB to examine the radiological consequences of disposing of blocks of vitrified high-level waste direct to the deep ocean.

392. These are two possibilities. A third, which may hold considerable promise, is to drill holes into stable areas of the bed beneath the deep ocean, emplace the vitrified cylinders, and backfill. We discuss these three options in more detail below, but we first mention briefly some other means of disposal that have been suggested but which appear fundamentally unsound.

These are:

(*a*) removal from the earth by rocket into space;

(*b*) subduction into the earth's mantle between the edges of tectonic plates in ocean trenches;

(*c*) deep burial in Antarctic or Greenland ice sheets.

393. The use of rockets to put payloads into orbit round the earth, or even to send them on more distant journeys is now well established. However, there is still a significant chance of failure, and the consequences of even one failure that resulted in the release of the wastes into the atmosphere would be so serious as to make the method quite unacceptable at present. Moreover, the energy required for disposal by rocket would be disproportionately high.

394. The earth's crust is now generally considered by geologists to consist of about a dozen plates which float on the liquid mantle and move relative to each other, albeit very slowly. At some plate boundaries, material from within the mantle emerges to form new areas of crust. At others, material from the leading edge of one plate is forced down under the trailing edge of the other and "subducted" back into the mantle. Plate boundaries are fairly well defined,

and Figure 16 shows where the main ones are believed to lie. All of them are regions of seismic activity in the form of earthquakes or volcanic eruptions. Some of the subduction zones are associated with the very deep trenches that exist in parts of the ocean. The proposal would be to place the blocks of vitrified waste either on or in the sea bed of such trenches, so that they would be drawn down into the mantle and permanently removed. The difficulty is that the time scale of the subduction process is fundamentally too long. It would probably take of the order of 250,000 years for the process to be complete; even plutonium–239 would have effectively decayed before this. Moreover, these areas are geologically unstable, and the blocks might be damaged while awaiting their final burial. The water is very deep—typically 10 km—and the technology for drilling in these depths has not been developed.

395. The ice sheets in Antarctica and covering most of Greenland are known to be of great antiquity—perhaps of the order of a million years. A proposal has been made([48]) that the vitrified canisters should be placed on the surface, where their heat would melt the ice under them. They would then sink down through the ice, which would refreeze above them. Eventually, after a delay which would be sufficient for the fission products (and curium) to decay to a significant extent, they would reach the bottom. But in 1973 it was suggested([49]) that underlying the ice sheets of Antarctica there were a number of lakes which are now believed to be in direct communication with each other and with the oceans. Thus the slow but inevitable grinding of the blocks by the ice on the basement rocks would lead to leaching and a pathway back to the biosphere for the actinides. Though the oceans would provide much dilution it seems unlikely that this proposal would be acceptable. Moreover, since operations in such inhospitable areas of the world would inevitably be both difficult and expensive, it might be simpler to provide the requisite period of delay by other means.

396. Of the three options that we regard as promising, disposal into the ocean directly would be much the cheapest. Moreover, there would be no need to provide interim cooling in an engineered facility except perhaps to reduce the hazards of handling very active blocks at sea and to provide buffer storage. But there are a number of uncertainties about the radiological consequences, and there is also the possibility of retrieval of the wastes by accident. At present such disposal would be prohibited under the London Convention, unless it could be conclusively demonstrated that the containment was so secure that the wastes would be effectively prevented from leaking into the ocean before the radioactivity was below the limits shown in Table 10. It appears unlikely that this could be satisfactorily guaranteed and we are very doubtful whether this method should be accepted as the preferred means of disposal on grounds of cheapness if other alternatives offering rather more convincing long-term security are available. The Institute of Oceanographic Sciences (IOS) in their evidence to us, with an admitted inbuilt bias in favour of preserving the ocean environment, argued that such a scheme was to be regarded as a decidedly inferior option as compared with emplacing the wastes beneath the ocean floor.

397. We have concluded that there are essentially two reasonable options for the permanent disposal of high level vitrified waste, namely geological

FIGURE 16

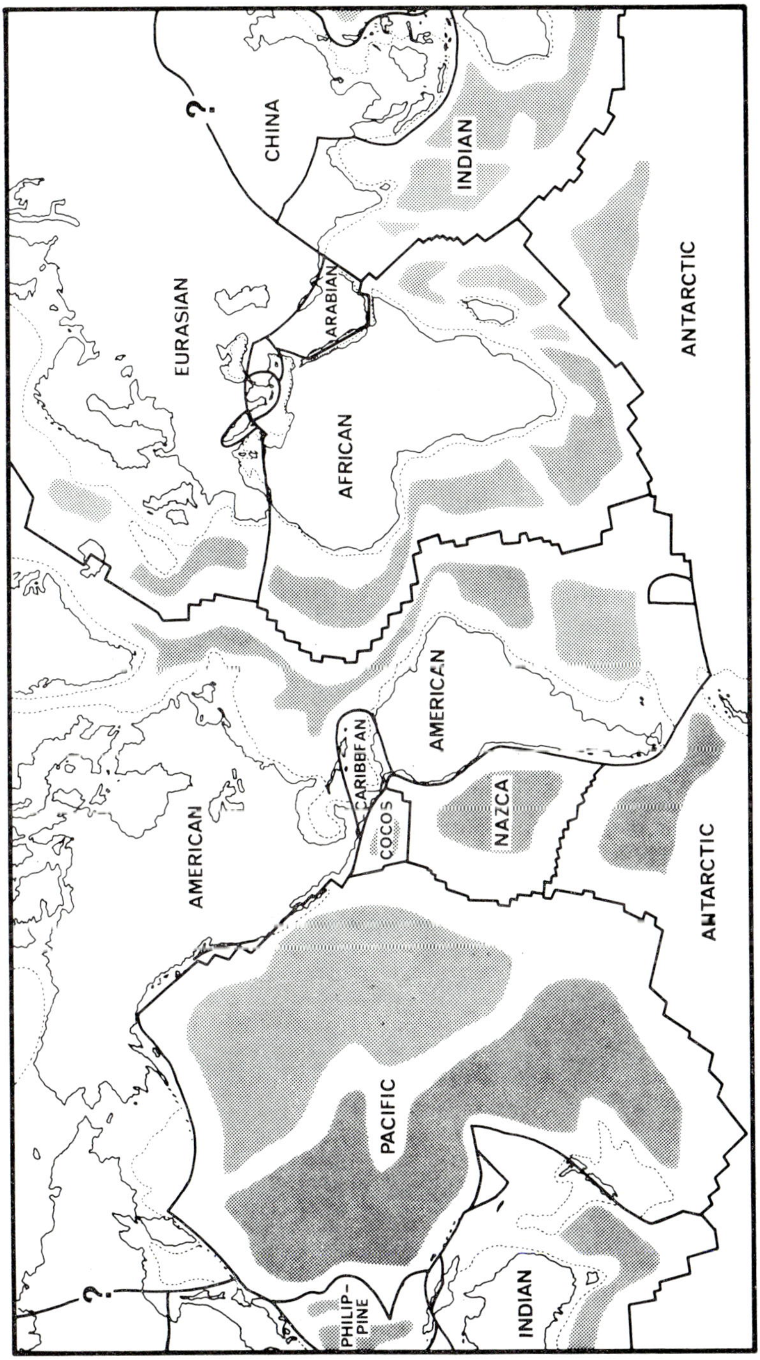

Institute of Oceanographic Sciences.

Map of the world showing boundaries of tectonic plates, and abyssal hill and swale regions (shaded areas).

formations on land and below the ocean floor in stable areas. Although we are not ourselves able to judge at this stage whether either of these options would be safe, we were sufficiently persuaded by the arguments advanced by the IGS and the IOS, who have no commitment to the promotion of fission power and who appeared to be expressing the consensus of views amoung their professional peers([50])([51]) that each offers a big enough chance of success to be worth pursuing seriously. We regard the demonstration of a safe disposal means for high level wastes to be an essential prerequisite to a big expansion of nuclear power. In our judgement therefore, work should be carried out vigorously in the UK on both options, preferably in collaboration with other countries, in order to assess their merits and probable relative costs. In the next two sections we shall describe the criteria that should be used to determine the two programmes of work, how they should be conducted and by whom, and the organisation needed to direct and pay for the research work and the subsequent disposal facility.

398. *Disposal to geological formations on land.* The basic information needed to allow an assessment of whether it would be practicable to develop a disposal facility in a given area is a reliable geological map which can be developed into a three-dimensional model. Next, the occurrence and movement of ground water within the area must be assessed, together with the physical properties of the host-rock. The actual construction of an appropriate excavation in which the wastes can be placed would depend on the development of appropriate mining, drilling and underground engineering techniques. We shall review the principles involved, but shall not attempt to identify areas within the UK that might be suitable.

399. The idea of disposing of long-lived high-level radioactive wastes in a geological formation that is physically, chemically and seismically stable has been advocated since the mid-1950s when such wastes were first generated in large amounts. There is the prospect of secure isolation from man's environment for very long periods of time, subject only to the possibility of transport in circulating ground water. Geologists, working on the "principle of uniformitarianism" under which existing geological processes are keys to the past and, by extrapolation, to the future as well, can have fair confidence that formations that have been stable for millions of years in the past will continue stable for periods of 100,000 years in the future. The principal uncertainties are likely to be the effect of climatic changes and alterations in sea level rather than in the solid geology.

400. The host-rock must not have fissures or intergranular pore spaces that would allow the passage of water and it must also be able to withstand the heat generated by the wastes (which could be sharply reduced by interim storage). Amongst other criteria, a disposal site should clearly be in a seismically quiet area and in this respect the UK is fortunate; as Figure 16 shows this country is well within the Eurasian plate. It should also be in an area of low or moderate relief, and remote from large bodies of surface water. The disposal facility should be at least 300 m below the surface to allow for possible erosion of the land surface under present and possible future climatic regimes. There

are three principal types of material which have been advocated—rock salt, which is favoured in West Germany and the USA for example; clay, which is being developed by the Belgians and Italians; and hard crystalline rocks, which are under investigation in Canada, France and Sweden.

401. Rock-salt exists both as horizontally bedded deposits (as in several areas in the UK, such as Cheshire, Somerset and Yorkshire) and as vertical masses extruded upwards by earth pressures into various shapes, described as "salt domes" or "diapirs", of which the Asse mine in Germany is an example. Salt is highly soluble in water, and its mere presence in thick formations indicates that these have been secure from ground water over long periods—generally tens of millions of years. It has further advantages, being easy to mine and a good conductor of heat. It tends to flow plastically, and cracks therefore are self-healing, but not so rapidly that excavated chambers cannot be kept open for adequate time for the emplacement of wastes. However, it is a valuable resource, and is therefore likely to attract mining companies. It was the late discovery of the presence of nearby solution mining that caused the US authorities to abandon the idea of using a salt formation near Lyons, Kansas as a Federal Repository for radioactive wastes. Investigations in a second formation in New Mexico have also encountered some brine, but are progressing. However, the Germans have fared better in Asse*, and they have several hundred more salt domes in the Federal Republic. There are none known in Britain, but a number are known to exist beneath the North Sea.

402. Clay formations have three distinct advantages. They have very low permeability, they can deform plastically to seal any gaps (including trial boreholes), and they can retard some potential pollutants such as radioisotopes by ion-exchange and adsorption. They are also able to cope with a source of heat; the intrusion of hot molten rocks into clay-like sediments is geologically quite common, and results in a narrow zone of baked clay around the intrusion. Clay deposits are of low economic value, and often quite thick. There are extensive areas in southern England and other areas of the UK that might be geologically quite suitable, and would be worth investigating as potential disposal sites, but there will undoubtedly be difficulties because of public reaction.

403. Hard rocks also comprise a wide range of formations. Since they were formed from the molten state, they are known to be able to withstand heat. In the mass they are impermeable, but being crystalline they have negligible plasticity and, at least close to the surface, are generally intersected by well-developed joints which could act as conduits for ground water derived from the surface. This is not always so at depth, however. The last occasion when there was any passage of mineralising solutions derived from depth in these formations in Britain was some 60 million years ago and that was confined to a few locations. There appears to be a negligible chance of early renewal of such activity in an area as stable tectonically as north-west Europe. Some recent

*However, the effects of a continuous heat source on the mechanical properties of the rock are not yet known with confidence.

research has shown that it is possible for the vitrified waste blocks to fuse themselves into the host-rocks through the heat flux from radioactive decay. Many of the remote districts of northern Scotland are composed of suitable hard crystalline rocks, including some of the offshore islands. A deep disposal facility on a small uninhabited island would be particularly advantageous if one were chosen that was separated hydrogeologically from the mainland. Any leakage of radioactivity into the island's ground water would be easily detected and in that event the dilution by sea water would provide a further line of defence.

404. Over the years 1959–1972, the AEA consulted the Institute of Geological Sciences (IGS) (formerly the Geological Survey of Great Britain) on a number of occasions with a view to the development of waste storage facilities on land. Most of the inquiries did not lead to any firm proposals, although as a result the Drigg site was approved for the disposal of very low level solid waste and, following geological investigations, disposal of wastes including both high levels of activity and long-lived isotopes continued in a shaft at Dounreay as mentioned above (paragraph 359). Some of the work was intended to provide a fall-back position in the event that sea disposal would not prove acceptable but, apart from these two facilities, no sites were selected for further work.

405. It was not until 1975 that the AEA commissioned the IGS to prepare a programme that would provide the underlying research and development needed to support a proposal to dispose of high-level waste. The programme was drawn up late that year, and envisaged ten distinct phases of work, requiring some twenty years before a disposal facility could be constructed. Most of the time would be needed for an adequate monitoring programme in order to provide confidence in the assessment of geological conditions, and for experience to be gained in the operation of a small pilot facility. This time-scale would be lengthened if there were delays in gaining access to the site for the necessary drilling work. The IGS estimated costs of the order of a few million pounds per site, and recommended that more than one should be investigated at the same time because otherwise a single failure might delay the whole effort by many years. Drilling represents the major part of the costs, and is significantly dearer in hard crystalline rocks than in clays.

406. These proposals have to be judged against the background of the international debates now taking place, for example those held under the auspices of the IAEA, which accept that geological disposal of high-level wastes is sound in principle, but that each formation and site require individual examination. It has been argued that regional facilities, accepting waste from several countries, would be desirable to take advantage of favourable geology, but we have seen no evidence that this principle would be accepted even within the context of the European Community. Although the UK now appears conspicuously backward among nations with significant nuclear programmes in its consideration and funding of studies related to geological disposal of radioactive waste, the time scale for such studies is necessarily long and no other

country has yet developed a working disposal facility for high-level waste. A major UK effort, building upon existing experience abroad, could make an important international contribution to the subject.

407. How might such an effort be implemented? At present the only source of the substantial funds that would be needed is the AEA, and the main body with the expertise needed to execute the work is the IGS. But the direction of the work requires a good appreciation of geological issues, and we are not persuaded that the AEA (or BNFL) possesses this. There could be conflicts of interest if the body best qualified to pronounce on the merits of a proposal were itself to be the executive agency for such a proposal under contract to a body without expertise of its own. Yet if the IGS were merely relegated to the role of advisor, not only might the proposal be developed without their expert guidance, but their own ability to judge might be weakened through lack of familiarity with the details.

408. We have no doubt that the IGS must do the work, but in such a way that they retain their independence of judgement. This can best be achieved if there is created within the IGS a small specialised unit which would require additional long-term funding from Government, perhaps by means of a 10-year contract to the Natural Environment Research Council (NERC). The work of the unit would need to be closely integrated with the appropriate sectors of the nuclear industry, and its outline programme approved by an advisory, multi-departmental, working party to whom it would formally report (see paragraph 429). It should publish all its results in the open literature. It should participate on behalf of the UK in all the international bodies that are concerned with geological disposal of radioactive waste, and it should also be represented on the national coordinating committees and working parties now being established.

409. *Disposal into the ocean bed.* This proposal, however sound it may be technically, has one disadvantage not possessed by disposal to geological formations on land, namely that the decision on whether it is acceptable may not lie only, or even principally, with the British Government. There may be objections from other states, particularly in view of the impending creation of an International Seabed Authority as a result of the conference on the Law of the Sea, held from 1974-76. Such a body may be greatly influenced by nations who fear the spread of nuclear power and hope to check it by placing difficulties in its path. On the other hand, the potential of suitable areas of ocean bed for the emplacement of high-level radioactive waste (given that the risks are shown to be negligible) may be seen as an asset, to be exploited for financial advantage. This might lead to greater international co-operation, and enable countries like Japan to dispose of their wastes with far greater safety than within their national boundaries.

410. Ocean bed disposal is likely to be more expensive than disposal on land, although such cost estimates as have been prepared, notably by the US Battelle North-West Laboratory(48), suggest that the capital requirements and

operating cost would be less than those of the associated fuel reprocessing plant, and would therefore have little effect on the economics of nuclear power. Even if the method is demonstrated to be significantly less hazardous than the disposal of wastes to a geological formation on land, its higher cost may make organisations with waste to dispose of chary of adopting it for an additional safety margin that may appear to them to be redundant. Clearly, until there are alternative schemes worked out in some detail it would be premature to judge, but at some stage a decision will have to be made, probably on an individual site basis. It should also be borne in mind that it is unlikely that ocean bed disposal will be a proposition for the very large volumes of intermediate activity solid waste which we have mentioned earlier: land burial may be the best solution here although possibly ocean dumping may be acceptable for a part of the total.

411. There are a number of potential advantages to the disposal of high-level wastes into stable areas of the ocean bed. In the first place, there are very extensive areas available, and some of them are of very low biological productivity and with no known mineral or hydrocarbon resources. These are two important criteria for site selection. Secondly, the wastes would be extremely secure from inadvertent retrieval, even in the far distant future if all records were lost. This could be important if it were decided that the dangers posed by large stocks of plutonium were socially unacceptable and it was decided to mix the plutonium (in oxide or nitrate form) with the high-level fission products so as to make it permanently inaccessible. Thirdly, some regions of the oceans are known to be among the most geologically stable areas of the planet, and the ocean bed under some 4,000 metres of water would not be affected by any conceivable climatic change or variation in sea level (perhaps 200 metres) caused by a complete melting of the ice caps on Greenland and Antarctica or a withdrawal of water caused by the onset of another ice age. However the bottom water itself, which comes from surface regions in high latitidues, would be altered. Fourthly, there are many layers of defence, all of which would have to be significantly breached before there was any danger to man. These include the glass itself and its container, any accretions on the cask surface, the impervious basalt or sediments in which it was embedded, the slow diffusion of isotopes through and retention on the overlying sediments by ion-exchange processes, and finally the very slow tidal currents in the bottom water. The only route back would be by way of the ocean, not through circulating groundwater which might be abstracted directly for drinking purposes.

412. The operation of a disposal facility in the deep ocean bed would, however, be a complex operation. Figure 17 shows in diagrammatic form what would be involved. The vitrified cylinders would be transported in shielded shipping casks, provided with means of cooling, to a ship, rather as irradiated fuel elements are taken from nuclear power stations abroad for reprocessing at Windscale. The ship would dock at a semi-submersible floating platform, where the vitrified cylinders would be removed from their casks and lowered into the prepared hole in the ocean bed. Such a hole might be up to 1,000 metres deep, and when full to the desired level, would be backfilled with an appropriate sealant to exclude circulating seawater.

FIGURE 17

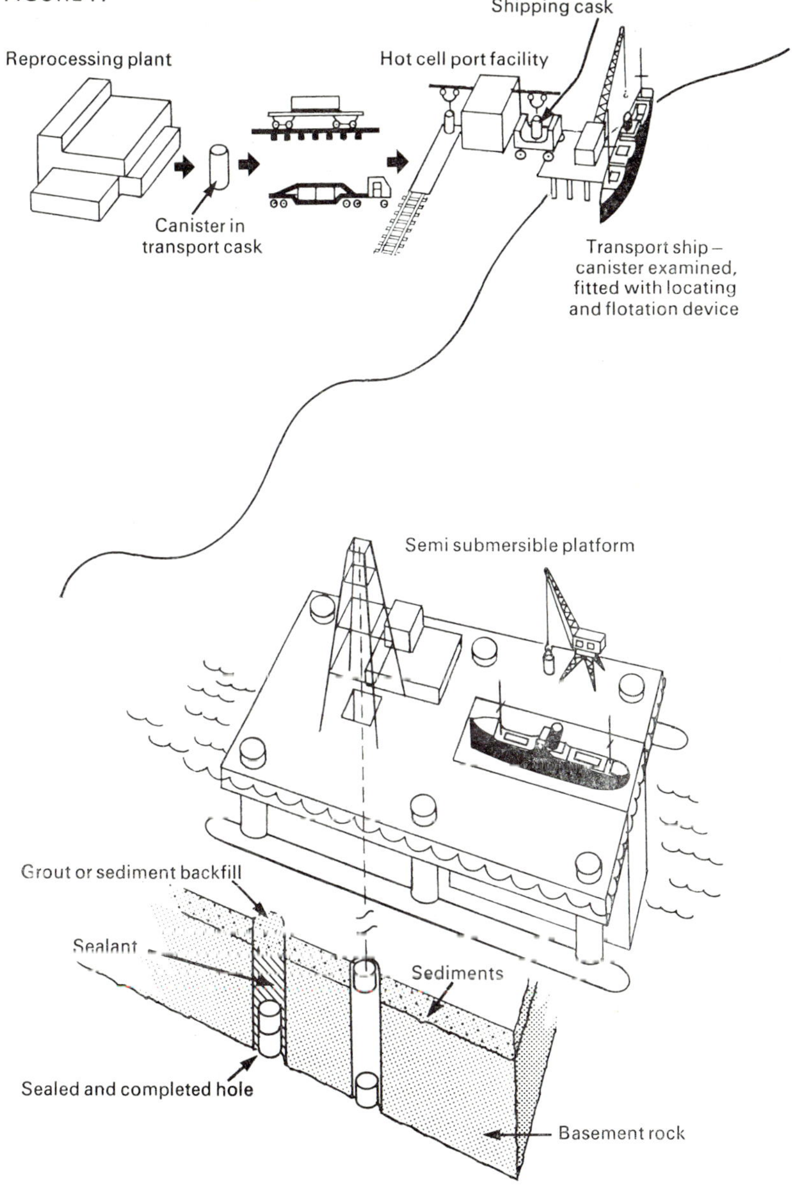

Diagram showing the major components of a system for disposing of radioactive waste in the ocean bed.

413. If the cylinders were 500mm diameter (as from the French plant at Marcoule), then a column 900m high would accommodate the fission product wastes from the generation of between 400 and 500 TWh of electricity, or some twice our total present annual consumption. With a total world installed nuclear capacity of 1200 GW, which some estimates assume might be deployed in 2000, it would be necessary to drill about 20 new holes a year. By way of comparison, 136 wells were begun in 1975 in the North Sea, and they were mostly several thousand metres deep. The Deep Sea Drilling Ship "Glomar Challenger" has already demonstrated its ability to drill holes up to 1,500 metres deep in 4,000 metres depth of water, and to replace its drill string in an existing hole. The cost is in the region of £10^6 per hole. Although certain improvements in drilling technology will be needed, there is little doubt that the impetus of searching for oil and gas in increasingly deep water will bring them forth. Thus neither the scale nor the technical difficulty of an ocean-bed disposal facility seem to raise insuperable objections.

414. Where might such a facility be located? The requirement for stability means that areas well away from the edges of the tectonic plates (Figure 16) should be chosen. The separation of plates and the formation of fresh areas of coastal surface takes place along the sinuous path of the mid-ocean ridge, a globe-circling rift some 40,000 km long. Spreading from this path there are numerous fissures which stretch across the flanks of the ridge and form wide fracture zones, in total extent perhaps one third of the area of ocean floor. Seamounts and remote islands occur here: they are examples of what can be achieved by volcanic forces, and are thus no advertisement for geological stability. Another requirement is that the area chosen should be relatively unproductive economically. This effectively rules out the continental margins where fish stocks are plentiful and hydrocarbon deposits are being exploited. These areas tend also to have unstable sediments and strong bottom currents which could return any escape of radioactivity to man.

415. The areas favoured by oceanographers are the abyssal hill and swale regions (undulating sediment-covered areas), which cover nearly a quarter of the surface of the globe including much of the Pacific Ocean and with smaller areas in the Atlantic and Indian Oceans, as Figure 16 shows. They are very gently sloping areas and have relatively few resources (except manganese nodules in some areas of the Pacific). Cores obtained by drilling in these regions tend to be very homogeneous, and seismic profiles and our growing understanding of deep ocean currents attest to the stability of these regions over tens of millions of years. There is, curiously, less information about their stability over the last tens of thousands of years because sediments deposited during this time are as yet unconsolidated, and the absence of evidence of earthquake activity would need to be checked more positively before they could safely be regarded as stable during the recent past as well.

416. These are two areas where more information is needed. Further work is required on the corrosion and leaching processes to which the waste cylinders would be subject in their abyssal holes, on the consequences of the heat plume

that would result, and on the transport processes in the basement rock, sediments and water column above. It seems unlikely that the heat from the waste will disturb the unconsolidated sediments and water mass above them very much, for not only will the water mass sweep back and forth twice a day with the tides, but the heat flow is likely to be smaller in magnitude than the variations known to occur naturally in different parts of the ocean. Much of the work could be conducted with "live" waste cylinders, for even if it were found that some of the processes were not entirely satisfactory, it is unlikely that any significant harm would be occasioned by the introduction of just a few vitrified blocks to such an environment.

417. At present, there is much less research on ocean-bed disposal than on disposal on dry land, and the main effort is in the United States, where the Sandia Laboratories in Albuquerque, New Mexico, have been managing a major programme on behalf of the Energy Research and Development Administration (ERDA)([51]). We have considered whether a major British effort should also be mounted and we have concluded that it should. In the first place, we believe that this option for the disposal of high-level waste really is a serious contender, and that it might well prove easier to implement than to find a site within the UK that did not arouse much local opposition. Secondly, we do not believe that it would be seen as duplication of the US effort if we were to investigate a site in the North Atlantic (which would, of course, be far more convenient for ourselves and other western European nations). If both the US work in the North Pacific and our own in the North Atlantic were to confirm the validity of the technical case, there would be a greater chance that it would gain international credibility. It would also be desirable to seek support from the European Community who might be expected to benefit from the work.

418. We have in mind a programme costing of the order of £500,000 a year, and which would be designed to enable a fair appreciation of the details of a disposal facility to be obtained in about ten years. We have been advised by the Institute of Oceanographic Sciences (IOS), whom we would expect to undertake the work on similar terms to the IGS, that these are reasonable targets. The programme, if successful, could lead to an operational disposal facility more quickly than could the land disposal project, as much of the engineering development work (on the cask, the ship, the semi-submersible rig, and the hole itself) could be carried out by different teams at the same time working on contracts administered by the IOS. This is an important advantage. Although we are advocating the work because it would protect the environment while allowing the British nuclear power industry to operate we are aware that the operation of a suitable disposal facility, together with the associated transport and reprocessing operations, could be an exceedingly good business, and one for which a technological maritime nation should have ambitions. As we mentioned earlier, this view is not at all dependent on our views on nuclear power development generally, for the waste exists now, is accumulating rapidly, and requires safe disposal. In fact, a shift away from nuclear fission would tend to make ocean-floor disposal more attractive

economically than land-based disposal, for the costs would tend to be proportional to the number of holes drilled rather than sunk in one large sub-surface facility where they would appear progressively smaller when spread over an increasing amount of waste.

Other issues in radioactive waste management

419. In the above paragraphs we have described the genesis of the various types of radioactive wastes and the management techniques currently employed, and we have discussed the areas where we believe that more work is needed. Before we go on to recommend the organisations that in our view are necessary to direct and execute the work, we raise a number of miscellaneous issues that have been debated from time to time and which we felt we should consider. There are three of them: the arrangements for transport of waste, foreign reprocessing business, and the likely effect of environmentally sound radioactive waste management practices on the cost of nuclear power.

420. *Transport.* The transport of radioactive materials, which are legally classed as "Dangerous Substances", is governed by regulations and controlled by the DOE's Dangerous Goods Branch if transport is by road or rail. The head of the branch is also an adviser in a personal capacity to the Department of Trade, who have responsibility for transport by sea and air. Containers for radioactive materials have to conform to standards laid down by the IAEA. "Class B" packages, which are used for the transport of irradiated fuel elements, have to be able to withstand a series of impact tests (representing the effects of a 30 mph crash) followed by half an hour in an intense fire. However we were surprised to learn that the tests are conducted only on models, and since the containers travel on ordinary freight trains which may be expected to travel at speeds up to twice that assumed in the tests (with kinetic energy four times that assumed), we were not wholly reassured. It seems unlikely that a container carrying Magnox fuel could give rise to more than a very local area of contamination even if it were to split open as a result of an accident occurring that was significantly more severe than that for which the flask had been designed. However the situation with respect to irradiated fast reactor fuel with its very much higher heat generation (see Figure 8) might be different, and we understand that the CEGB and British Rail are in fact collaborating on the design of a flask that could withstand even a direct blow from a tunnel or bridge abutment at the maximum speed of the train.

421. We formed the impression that, although the present practices did not appear to give rise to a significant public hazard, this might not be so in future when the numbers of flask movements may well be much greater, and active cooling has to be provided even for fuel travelling within the UK. (At present the flask fins are adequate to dissipate the heat). There seem to be a number of issues requiring review, such as whether all future nuclear power stations should be provided with a railhead to avoid transfers from road to rail, whether underwater loading (which at present leads to a certain amount of difficulty with decontamination) should cease, and whether there should be regular checks or metallurgical examinations of fuel flasks. We have concluded that

since a flask is in effect a small-scale nuclear installation, with fuel, moderator, and coolant, the transport of irradiated fuel should formally require the approval of the Health and Safety Executive, who would presumably base their licence upon the technical advice of the Nuclear Installations Inspectorate. There should be a system of regular inspections of the flasks, including those imported from abroad. These are at present monitored upon arrival by BNFL staff; we consider that there should also be some checks by DOE as guardians of the environment.

422. *Foreign reprocessing work.* The question of whether BNFL should import substantial quantities of irradiated nuclear fuel for reprocessing from Japan became a live political issue at the end of 1975. There was a somewhat inconclusive public debate in January of this year in London, and on 12 March 1976 the Secretary of State for Energy announced in Parliament that the company could take on foreign business but on the condition that the contract should give the company an option to ship back the residual radioactive waste. Nothing was said about what would happen to the plutonium, or the conditions under which it would be returned to the country of origin, which was one of the principal objections raised to the contract in the public debate. We have already discussed this issue in paragraph 319.

423. From the point of view of radioactive waste management and its effect on the environment, we cannot see any grounds for objection to the contract, but we would not think it desirable that the option described by the Secretary of State should be exercised. The transport back to the country of origin of radioactive wastes (or the return of irradiated fuel in the event that oxide fuel reprocessing or the vitrification process are not successfully developed) would pose risks which would be best avoided, and the eventual safe storage of the wastes might well be much better assured if they were retained in this country. This is particularly likely to be so in relation to a country like Japan which is notably prone to seismic activity.

424. In our view, therefore, if the British Government is prepared to sanction foreign reprocessing work for commercial reasons, it should also be prepared to accept ultimate responsibility for all the material, once it has been received. In any event we shall have to devise satisfactory means for the safe containment or disposal of either unreprocessed fuel or high level wastes from our own reactors.

425. *The costs of sound radioactive waste management.* In many areas where pollution from industry poses a threat to human health or to the environment, the process of cleaning up is less thorough than we might wish because of the high costs involved. Sometimes the argument is heard that the costs of maintaining a clean environment would be so high that the industry would be uneconomic and would be forced to close down, with consequent damage to the economy. During our study into nuclear power this argument was never raised. Indeed there appears to be a remarkable degree of acceptance by the industry that it has a duty to do what is needed in order to ensure that the

levels of radioactivity in the environment are not significantly increased as a result of waste discharges. The Department of Energy in their evidence accepted that each energy system should "bear the additional costs . . . necessary to make it environmentally acceptable". We agree, although the question is begged who should judge the acceptability of a practice. In our view, this must be a department of state whose function is to protect the environment.

426. We make no apology, therefore, for advocating courses of action that would add to the expense of nuclear power. However, such cost figures as we have been able to glean, mainly from Environmental Impact Statements issued by the Energy Research and Development Administration, ERDA (formerly the Atomic Energy Commission, AEC) in the USA([52])([53]) do not suggest that these costs would make nuclear power significantly dearer. In 1974, the AEC estimated that it might cost $3 billion (say £2,000 million in 1976 £) to treat, store and finally dispose of all the radioactive waste generated in the course of producing 60,000 TWh of electrical energy up to year 2000. This represents a cost of about 0·0033p per kWh, or less than 1 per cent of the cost of generating electricity in Magnox reactors. Although this figure may well be an underestimate as there are a number of technical problems to which the answers are not yet known, we cannot believe that sound waste management practices should be handicapped by the costs involved, whether these be for execution or for underlying research and development.

The future: policy and its execution

427. The picture that emerges from our review of radioactive waste management is in many ways a disquieting one, indicating insufficient appreciation of long-term requirements either by government departments or by other organisations concerned. In view of the long lead times that will almost certainly be involved in the development of appropriate disposal facilities, we are convinced that a much more urgent approach is needed, and that responsibilities for devising policy and for executing it need to be more clearly assigned.

428. We are clear that the responsibility for developing the best strategy for dealing with radioactive wastes is one for the Government, and specifically for a department concerned to protect the environment, not one concerned to promote nuclear power. It must therefore fall to the Secretary of State for the Environment. We recommend that there should be established a Nuclear Waste Management Advisory Committee to advise the Secretary of State; this should be a statutory body which would replace the recently created Radioactive Waste Management Co-ordinating Committee. The Advisory Committee should have a strong environmental representation. We envisage that the Chairman and a majority of its members should be scientists appointed by the Secretary of State after consultation with the Minister for Agriculture, Fisheries and Food and with the NERC. It should also have representatives from the DOE and MAFF, from the Department of Energy, from HMPI (if that body is set up, otherwise from the RCI), from the ARC, and the NRPB. Industrial representatives should also be appointed from the AEA, BNFL, the utilities, and the executive organisation for waste disposal that we discuss below.

429. The Advisory Committee would be concerned only with broad policy objectives. We envisage that, since it would need to see the results of appropriate development work before it could proffer informed advice, it would have to have funds guaranteed for long enough to allow it to commission long-term (10–20 year) studies. It would, in fact, be the natural body to sponsor and direct the geological and oceanographic studies we have outlined above. Subsequently, when suitable disposal facilities were in operation, it would have a less active role to play. We expect that, in giving advice to the Secretary of State, it would report on any technical developments which might exacerbate disposal problems, so that these could be fully considered before they were allowed to proceed. Conversely, it would be expected to indicate the direction of future development work that could be expected to reduce the problems of waste disposal.

430. There is also a need for an executive organisation that can develop and manage radioactive waste disposal facilities, and accept solid waste from those who create it. Although it would have much to contribute to the formation of national policy on waste management, its primary function would be an operational one, and it would act rather like the companies who accept waste containing toxic chemicals and dispose of it or render it harmless for a fee. We think, however, that the aim of the organisation should be to protect the environment rather than to offer the cheapest terms to BNFL, the electrical utilities and others who might require its services. It would therefore have to have a statutory monopoly, and it should be accountable to government, specifically to the Secretary of State for the Environment, for its activities.

431. We do not think that there is any existing organisation that could be given such a task. The institutes of NERC will be involved in the initial development of pilot facilities as a part of their general research into land-based or ocean-bed disposal options. They may well continue to monitor the activities of a disposal facility when it is in full-scale operation, but clearly they would not be appropriate bodies for the business of disposing of wastes as a commercial proposition. We do not think either that the CEGB or BNFL could conveniently have their remit extended to embrace waste disposal. It would inevitably be a subsidiary activity, and it is hard to see that part of either organisation could be made separately accountable to the Secretary of State for the Environment; certainly there might be awkward conflicts of interest.

432. We have therefore reached the conclusion that a Nuclear Waste Disposal Corporation (NWDC) should be created. Its function would be to develop one or more disposal facilities, and there dispose of all solid radioactive waste, packaged in accordance with its requirements, to the satisfaction of the DOE (if on land) and of MAFF (if at sea or in the ocean bed). It would follow guidelines laid down by DOE on the advice of the Advisory Committee, and would publish an annual account of its activities. The Secretary of State for the Environment should nominate the Chairman and Board of Directors, in consultation with MAFF and the NERC, and there should be directors representing the AEA, BNFL, the CEGB, the SSEB and the NRPB.

433. Through the charges levied by the NWDC, the nuclear industry would be seen to meet the whole cost of rendering its wastes environmentally acceptable. This is in accord with the principle accepted by the Department of Energy for comparisons between different energy options, that each should bear the additional costs that would be necessary to protect the environment. We envisage that the NWDC would also be able to seek and accept foreign radioactive waste, either material reprocessed by BNFL, or waste from foreign reprocessors or utilities. We have referred earlier (paragraph 418) to the possibility that this might prove to be a remunerative venture.

434. All those organisations generating radioactive waste would have to satisfy DOE that they were able to dispose of it to the environment within their permitted authorisation, or were packaging it for disposal by the NWDC. Some temporary storage of waste on-site would clearly still be required in order to allow short-lived radioisotopes to decay, but all nuclear installations would have a time limit for such storage, with a requirement that it be taken over by the NWDC in accordance with the latter's requirements.

435. We have thus recommended two new organisations, to devise waste management policy and to implement sound disposal practices, and that each should be responsible to the Secretary of State for the Environment. In so recommending, we are consciously placing heavy responsibilities on the DOE, who will have to take steps if need be to equip themselves with the necessary expertise. Until now, there has been a diffuse pattern of responsibility with the result that too little has been done. This will not do in future. Radioactive waste management is a profoundly serious issue, central to the environmental evaluation of a nuclear power programme. There must be a clear, identifiable, policy centre and a means to ensure that the issues posed by waste management are fully considered at the outset of a nuclear programme, not dealt with many years after the decisions on developments that lead to the wastes have been made and when options may have been effectively foreclosed.

CHAPTER IX

ENERGY STRATEGY AND THE ENVIRONMENT

Introduction

436. In this Report we have considered the environmental implications of a large nuclear power programme. Whether these implications should, or must, be accepted depends in the last resort on the extent to which such a programme fits into the choices made, and this raises the question of the alternative strategies that might be available and of their economic, social and environmental consequences. We are not an energy commission and we are not in a position to attempt a full analysis of the many factors that are relevant to the determination of energy policy. Nevertheless, we decided that our Report would be incomplete if we were not to seek, and attempt to provide, some understanding of those issues that bear on the question of whether great future dependence on nuclear fission power must be regarded as inevitable. That is our aim in this chapter.

Technical considerations

437. *Units.* In order to discuss energy questions some basic technical matters need to be appreciated. There is first the question of the units in which energy is to be expressed. A variety of units is used in different publications, making comparisons difficult for non-technical readers. Under the international system (SI) energy is measured in joules, but we shall take as our basic unit one kilowatt-hour (kWh) which is familar to users of electricity as the "unit" of supply. (One kilowatt-hour is equivallent to 3·6 million joules). This is, for example, the amount of electrical energy that would be used by a one kilowatt electric fire in one hour. However, 1kWh is far too small a unit to be convenient for expressing energy usage on a national scale and we shall use a multiple, 10^9kWh or one terawatt-hour (TWh), throughout this chapter.

438. Though the kilowatt-hour is familar only in relation to electricity it can also be used to measure energy in other forms. Some domestic heating appliances are now being rated in kW even though they burn gas, oil or solid fuel. Thus the use of TWh as a unit can be applied equally to combustible fuels which are used to provide heat. The Table below gives equivalent values, on the assumption of 100 per cent efficient combustion*. In practice, of course, any appliance will be less than 100 per cent efficient, as the combustion may not be complete and some of the heat produced inevitably accompanies the hot exhaust gases into the atmosphere.

*The values given are those normally used for conversion purposes, and are based on the assumption that the water formed by combustion of hydrogen in the fuel is not condensed.

TABLE 11
Approximate energy equivalents of 1TWh

Heat units	34·12 million therms = 3·412 × 10^{12}BTU 860 billion kilocalories = 860TCal 3·6 × 10^{15} joules = 3·6 million gigajoules (GJ)
Coal	124,000 tonnes (average figure world wide) 135,000 tonnes (UK in 1974)
Oil	83,000 tonnes
Natural gas	96 million cubic metres = 3·4 billion cubic feet
Hydro-power	3·67 billion tonnes of water falling 100 metres
Uranium	9·8 tonnes at utilisation of ½% (thermal reactor range) 65kg at utilisation of 75% (breeder reactor range)

439. The Table illustrates the relative compactness of different forms of potential energy. Thus falling water must be used in enormous quantities to provide 1TWh. The tonnages of uranium required to produce an equivalent amount of energy are, by contrast, extremely small. They may however require the removal of very large amounts of ore with consequent environmental degradation. For example, if a thermal reactor, fissioning only 0·5 per cent of the nuclei present in the uranium were to run on uranium produced from Chattanooga Shale (0·006 per cent uranium), some 163,000 tonnes of ore would have to be processed to yield 1TWh of heat, almost as much as the amount of coal mining required (assuming minestone of about half the weight of coal produced). It would however be easier to transport.

440. *Efficiency and conversion.* Energy is used for various purposes by consumers, but in order to reach the point of use "primary energy" will usually have to be converted to "secondary fuel" by the energy industries. These use energy in the process, and further losses occur in the distribution of fuels to consumers. The energy is finally used in appliances which are themselves less than 100 per cent efficient. The losses involved in energy conversion and distribution are well established and detailed statistics are available, but there is much less information on the efficiency with which energy is used by final consumers.

441. The conversion of heat energy into mechanical energy is of particular importance. This is the process by which the heat energy available in coal, oil or nuclear fuel is transformed into electricity. The conversion is subject to the second law of thermodynamics which dictates that the efficiency theoretically attainable depends upon the maximum and minimum temperatures of the working fluid, usually steam. These temperatures are limited by engineering considerations and in practice the overall efficiency even of the most efficient power station is barely 35 per cent. The average is about 30 per cent. This means that in practice only about one-third of the heat energy latent in the fuel (the primary energy) is released as electricity. Generally the rest of the heat energy is discharged to the environment, through the medium of cooling water, as waste, low-grade (i.e. low temperature) heat. The further losses involved in the transmission and distribution of electricity from power stations to customers must be taken into account to arrive at the overall efficiency. In 1975, for example, these losses amounted to 10 per cent, and the overall efficiency was 27 per cent, for the electricity industry in the UK([54]).

442. Smaller energy losses or overheads occur in other energy conversion and distribution processes. Thus, the refining of crude oil into saleable products, the mining of coal and its conversion into other solid fuels, and the distribution of natural gas by pipeline involve losses, respectively, of about 8 per cent, 2 per cent and 6 per cent of the primary fuel converted or produced. In the future, however, it may be necessary to manufacture synthetic natural gas (SNG) from coal, and for this process the losses are likely to be more substantial, of the order of 35 per cent.

443. *Consumer requirements.* Energy is needed for four main purposes—low temperature heat (to heat buildings), high temperature heat (for industrial processes or, in the home, for cooking), mechanical power and specifically electrical applications (such as lighting and electronic appliances). Energy for transport, except for electric railways, needs to be in the form of portable fuels. Electricity can be converted into mechanical power with relatively small losses and is thus convenient to use for this purpose in fixed installations. Alternatives such as the internal combustion engine have mechanical efficiencies of typically only 25per cent and so suffer losses comparable to those of electricity generation.

444. Because of the high losses involved in electricity generation it is inherently wasteful to use electricity as a source of heat except in those particular applications where its controllability offers a compensating advantage. It is also wasteful, however, in thermodynamic terms, to burn a fossil fuel, such as gas or oil, at a high temperature and then to degrade this heat for domestic heating or hot water supply. A determined approach to efficiency in the use of energy would seek to match the types of energy source to different user needs and would probably involve the distribution of hot water, gas and electricity to buildings.

445. Increases in the efficiency of energy use can be obtained through district heating schemes in which the heating load for a district is supplied from a large, central boiler plant by piping hot water to the houses or other premises connected to the system. Such schemes are applicable only to urban areas where the load is sufficient to justify the capital cost of the distribution network. The advantage lies in the fact that the central plant can be operated at higher efficiency than individual appliances, which more than compensates for heat losses in distribution. Moreover, the boiler plant can more readily be adapted to burn a range of fuels, including fuels of lower grade. There are other gains such as reduced maintenance, more effective pollution control and reduced fuel transport requirements. There are only a few district heating schemes in the UK in contrast to many in some other European countries.

446. Still further gains in efficiency can be achieved by the use of combined heat and electrical power (CHP) systems which offer a means of matching the type of energy supplied to the users' needs. In such systems the central boiler plant referred to above is replaced by a power station. It is perfectly feasible to operate an electrical generating plant so that the "waste" heat now discarded to the environment is produced at a higher temperature suitable for district heating purposes. The electrical energy produced for a given input of primary

energy is of course reduced, but since both electrical and heat outputs can now be used the overall efficiency of the system is much higher and can be in the range of 70–80 per cent. Combined heat and power systems of this kind are again applicable only in urban areas and they require the siting of the power stations within about 30 km of these areas in order to achieve acceptable lengths for the hot water distribution network. Again, systems of this kind have been introduced on only a very limited scale in the UK but are in widespread use in other countries (see paragraph 491).

447. Energy supply strategy must also take account of the variation in demand for the different kinds of energy at various times of day and seasons of the year. Figure 18 shows the demand for electricity on the CEGB grid on two extreme days in 1974: there is a variation in load of two-to-one within the day and between winter and summer extremes. A similar pattern would be seen for gas and other fuels. Sources of energy differ in the ease with which they can be stored. At present electricity can be stored in bulk only in the form of water in an elevated reservoir, which releases its potential energy when it flows down through turbines to a lower reservoir. There are losses in the process, and moreover the creation of artificial reservoirs often raises environmental objection. Gas can be stored to cover diurnal fluctuations in demand, but to accommodate seasonal changes it may be necessary increasingly to boost natural gas supplies with SNG in winter time. Coal and oil, in contrast, are easily stored to cover seasonal variations in demand, though at some expense.

448. *Energy supply and pollution.* Energy conversion and use generally cause pollution, and energy conservation, in the sense of obtaining the maximum utility from each unit of energy purchased, will not only ease the supply position but also reduce pollution because of the reduced requirement for primary energy. Thus, the open coal fire has an efficiency of only about 30 per cent and its replacement by gas or oil-fired heating systems not only raised the efficiency to 60–70 per cent but has brought about a dramatic reduction in the amount of smoke in urban atmospheres. Pollution control implemented under the Clean Air Act thus made a notable contribution to the efficiency of energy use. Electric heating has also increased since the time of this Act and has contributed to the cleanliness of cities. The efficiency of conversion of electricity to heat is high (though for night storage heaters it is only about 70 per cent since some of the heat is given out when it is not required([55])), but because of the low overall efficiency of electricity generation, electric heating, compared with gas or oil, generally results in higher emissions of sulphur dioxide. This pollution is transferred to the power stations where it is widely dispersed through tall stacks. Some of it travels as far as Scandinavia where it contributes to "acid rain"([56]).

449. We consider some aspects of pollution arising from energy conversion later in this chapter, but we note here that pollution is also associated with the distribution of energy. Electricity is transmitted over a large (and expensive) network of pylons and cables which are intrusive and unsightly. Opposition to a new power station is often just as much to the associated wayleave needed for the transmission line as to the station itself, which may in fact be relatively

acceptable in visual terms (see Plate 1). Gas is distributed by pipelines underground which, once installed, cause no environmental problems apart from occasional leaks. Both coal and oil are distributed by rail and by road, and heavy vehicles contribute their share to the deteriorating urban environment.

FIGURE 18

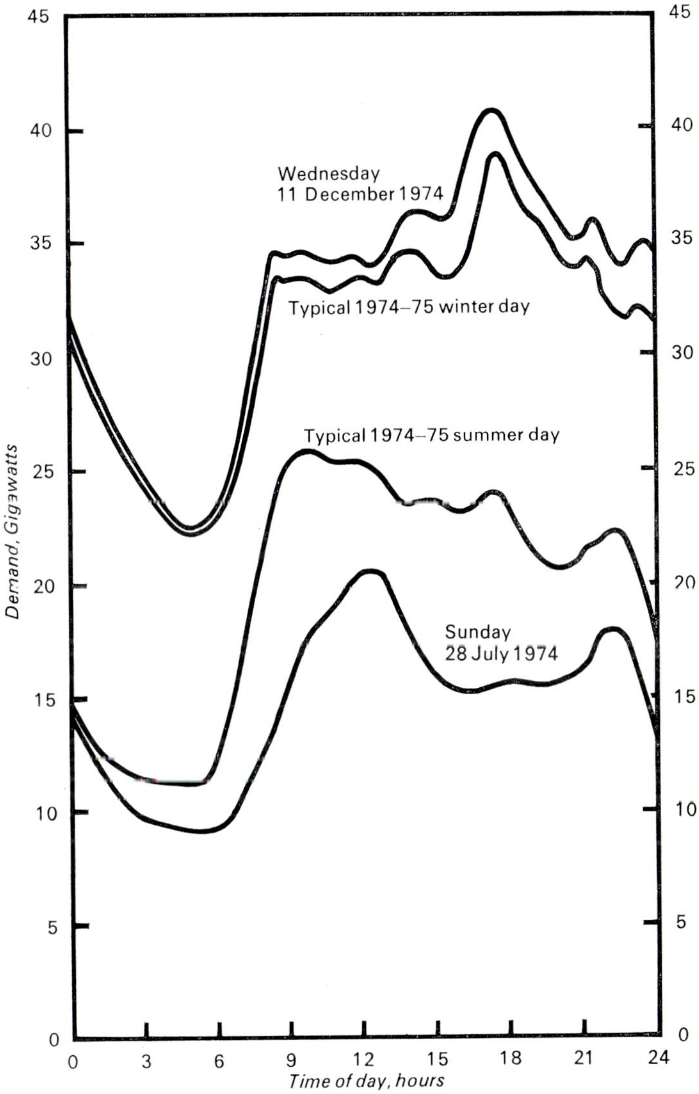

Summer and winter demands on CEGB system in 1974/5 including days of maximum and minimum demands

Central Electricity Generating Board.

Chapter IX

World energy demand and supply

450. In 1975 the world consumption of primary energy was about 75,000 TWh(57) of which the UK, with 1·4 per cent of the population, used about 3·3 per cent. Consumption has been expanding at about 5 per cent per annum since 1960 and with world population increasing at about 2 per cent per annum it is expected to continue to grow, though possibly at a lower rate in future in view of the recent dramatic increase in the price of oil. Of the total primary consumption, some 43,300 TWh or 58 per cent was used by OECD nations, 22,000 TWh or 29 per cent by eastern Europe, the USSR and China and only 9,700 TWh or 13 per cent by the rest of the world. The dominant component of the total growth in demand has been that of the industrial nations. If their consumption were to stabilise, very large percentage increases could occur in the amount of energy used by developing nations without a major effect on projected world demand.

451. On the basis that increased economic activity depends upon increased energy use and that there is a positive correlation between national energy use per head and GNP, forecasts have been made which suggest that demand for energy will continue to rise rapidly in order to satisfy economic aspirations. However, examination of Figure 19 suggests that this need not necessarily be so; energy demand depends upon the industrial structure, transport policy, fuel prices, advertising, attitudes to conservation and a host of other factors. Nations vary greatly in their use of primary energy per unit of GNP. Britain has one of the highest ratios among western industrial nations and is in the same bracket as countries like Canada and Norway which have abundant hydro-electric power and have therefore developed industries that depend upon it.

452. The expansion in world demand for energy over the last few decades has been met by rapidly expanding supplies of fossil fuels, mainly oil and natural gas. Their price in real terms dropped during the 1950s and 1960s with the discovery and development of the massive reserves in the Middle East. Further oil reserves are continually being discovered, but in increasingly hostile environments such as the North Sea and Alaska, and their exploitation will involve dangers both to the operators and the natural world. The actual extent of the world's recoverable oil reserves is difficult to assess, but seems likely to be of the order of 230×10^9 tonnes, equivalent to some 2·8 million TWh of energy. This is over 80 years' supply at the current rate of oil consumption (33,000 TWh in 1974), but during the last decade this has been increasing at about 7 per cent per annum and reserves would last only until about the end of the century if expansion continued at the same rate. In fact increasing price will tend to limit consumption and enforce major changes in consumption patterns.

453. There has been a similarly rapid expansion in the use of natural gas, but the annual consumption, 13,500 TWh in 1975, is less than half that of oil so that the estimated total reserves of 2·1 million TWh should last rather longer—over 150 years at the present consumption rate but only until 2010 if the current rate of increase in consumption was to continue unchecked. Coal reserves are much less well defined, mainly because there is doubt about what fraction of the

total resources thought to exist—between 70 and 80 million TWh equivalent—are economically recoverable. This could be around 4·2 million TWh at present energy prices and with currently available techniques, but both may increase in the future and some observers foresee a developing world trade in coal towards the end of the century similar in scale to the oil trade that exists now. At the current rate of consumption of about 23,000 TWh per year, the ultimate resources might last several thousand years though currently available reserves would suffice for only about 200 years. However, as oil and gas are successively worked out, coal will be needed to make portable fuels and SNG, and its consumption may be expected to increase more rapidly than at present.

FIGURE 19

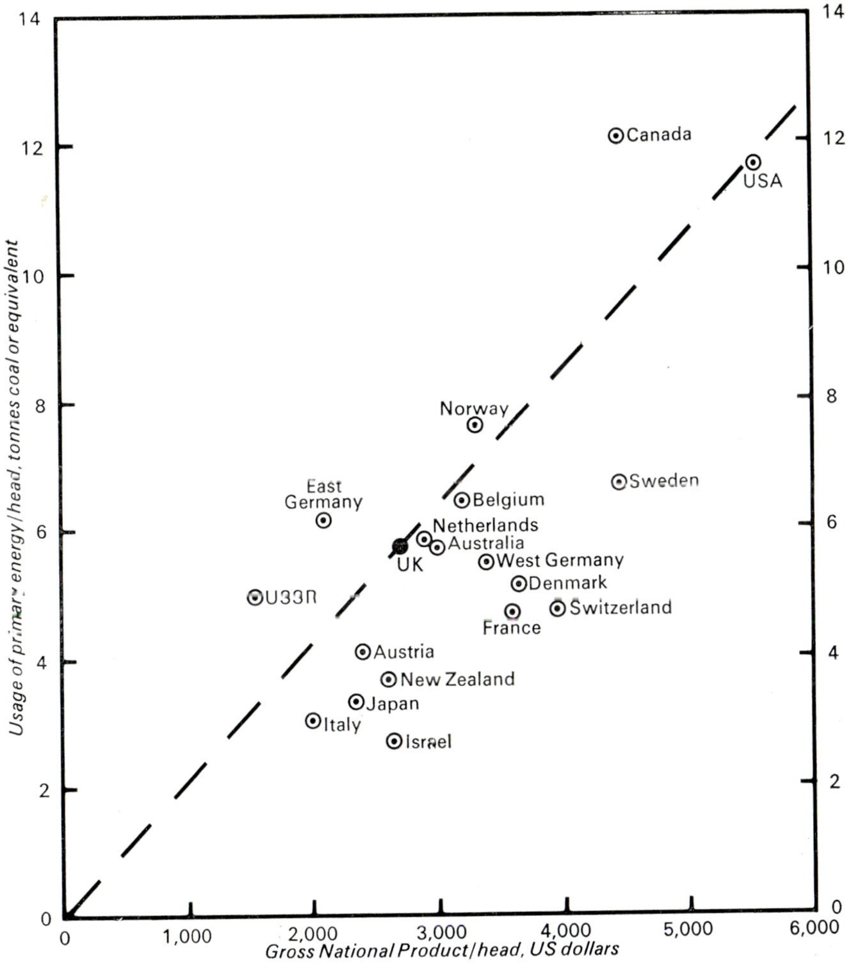

Usage of energy and gross national product per head for selected industrial countries in 1972.

454. This brief review of fossil fuel supplies suggests that the limitations of a finite supply and an exponentially increasing demand (if it continues) could lead to severe shortages in the next few decades unless the industrially developed nations curb their growing demands for energy rather sharply. The "gap" between supply and demand is illustrated in Figure 20. If the recent trend in the growth of energy consumption is extrapolated into the future, the "gap" between supply and demand will be huge and could only be met by the development of major alternative sources of energy.

455. There appears to be general agreement that nuclear fission power is the only alternative source of energy now available that is at a sufficiently mature stage of development to be capable of making a significant contribution to meeting such a "gap" on the time scales that apply. It has been suggested([58]) that one half of the probable "gap" between world energy supply and demand early in the next century could be met by nuclear power if some 3,500 GW of nuclear plant were in operation by year 2000, and that the rate of reactor building this would imply would be possible. Such a programme however would call for the commissioning throughout the world of nearly three large (1,000 MW) reactors each week on average for the rest of the century.

456. Uranium reserves might well be insufficient for such a programme if it was restricted to thermal reactors. World uranium resources are known to be very great but the amount that exists in concentrations high enough to be economically workable is not known with any precision as large areas are not explored. Even if suitable reserves are discovered, it can take more than a decade to bring them into production, and thus there are fears of a severe uranium shortage in the future. Since there is already a large number of nuclear reactors in operation (total capacity about 70 GW at the end of 1975) and under construction (a further 157 GW expected to come into service by 1980), the forward price of uranium has risen sharply in recent years. Uranium reserves are quoted in terms of a price at which recovery would be economic—the higher the price the greater the reserves. A 1973 estimate([59]) suggested that a price of US$50/lb (which is about twice the current price) would allow extraction of some 6·9 million tonnes of uranium, equivalent to the generation of about 225,000 TWh of electricity in thermal reactors. For a world nuclear programme of 3,500 GW this amount of uranium would last only about 12 years.

457. The same quantity of uranium used in fast breeder reactors would produce about 42 million TWh of electricity, an amount of energy many times as great as that available from all fossil fuel reserves under current conditions. However, as we described in paragraph 113, FBRs depend on an initial supply of plutonium which has to be created in thermal reactors and the rate at which the former could be introduced in the early stages of a world nuclear programme would depend on the rate at which thermal reactors could be built and fuelled. It has been estimated that constraints on uranium supply would be likely to limit the number of thermal reactors in service by 2000 to about 800 GW and

FIGURE 20

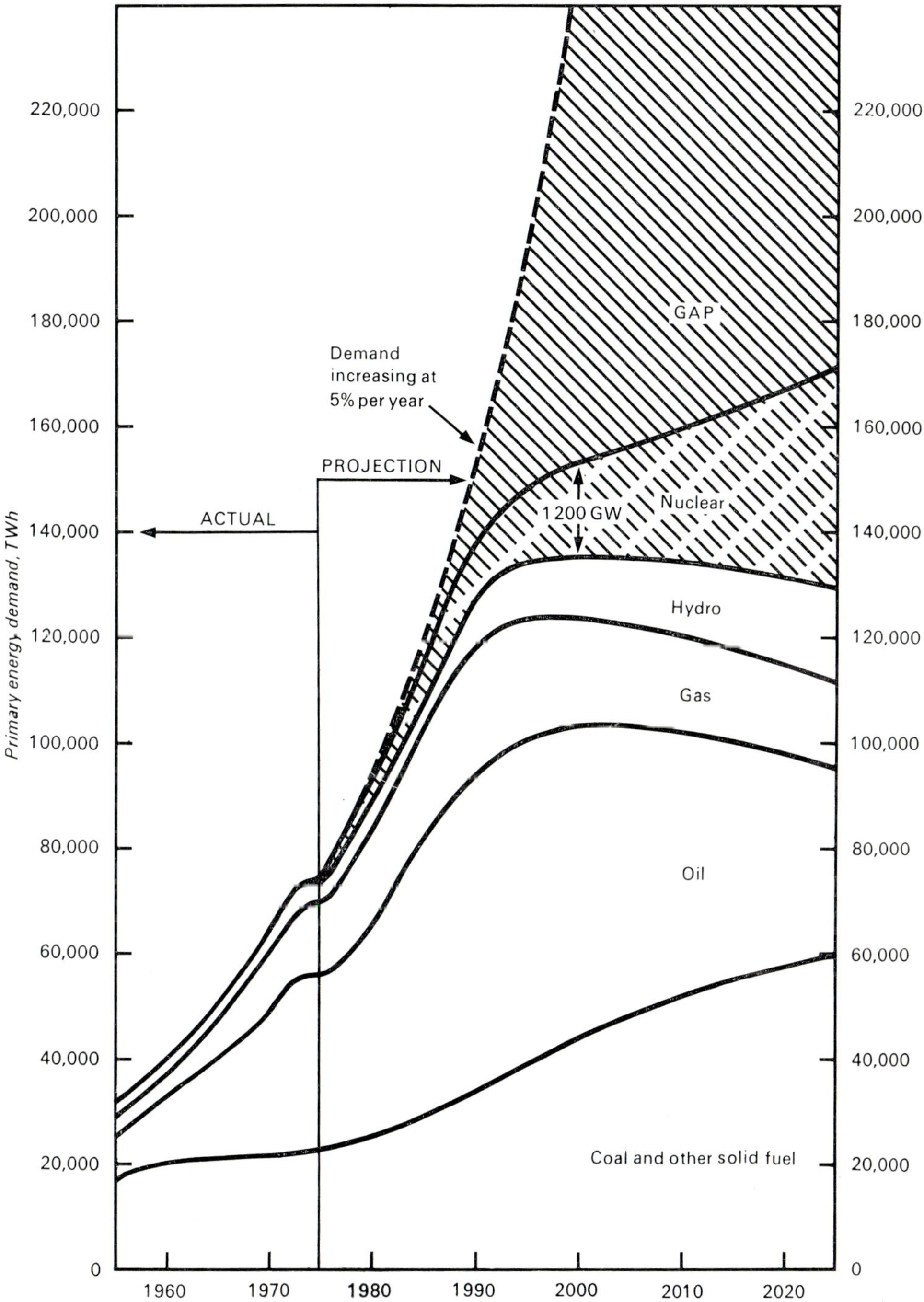

World energy demand, and a possible means of supply, projected to 2025

that, given an early start to FBR use, these could increase capacity at that time by about 50 per cent. On this basis nuclear capacity early in the next century would be sufficient to meet only about one-sixth of the world energy "gap"; the logic of this argument points to the conclusion that the "gap" cannot be filled, certainly not by nuclear power alone. In that case there would be increasingly severe constraints on expansion in demand and world energy prices would rise steadily in real terms to enforce economies.

458. Apart from nuclear fission power we can at present look only to the so-called "natural" sources as additional sources of energy supply. We should include in this term hydro-electric power (which at present supplies about 6 per cent of world primary energy) and the extensive use of firewood in many under-developed countries. Both of these can have substantial environmental effects; in particular the cutting down of forests may have drastic consequences such as soil erosion, the creation of deserts and adverse climatic changes.

459. Other natural sources include the direct use of sunlight as heat or its conversion to electricity, and its indirect effects in forming wind and waves. In a few favoured areas there are also possibilities for the use of geothermal heat or for the exploitation of the energy in the tides, but neither is likely to make more than a local contribution. The exploitation of these sources may have substantial environmental effects as has been shown in New Zealand([60]) and suggested for a tidal barrage on the Severn estuary([61]). Sunlight and its indirect forms are potentially of sufficient magnitude to provide for world energy needs at a much higher level even than now. For example, the total solar energy received by the UK alone is three times the present total world energy consumption. However it is diffuse, and its availability is subject to the vagaries of the weather (although generally wind and wave energy in the UK are greater in the winter than summer so that, in average terms, the energy supply would correspond to the pattern of demand). However the technology needed to exploit solar energy in its different forms has not been intensively developed while cheap and plentiful supplies of fossil fuels were available. On the other hand, many years are generally needed for such developments, so that they may not be available when they are needed. This is a basic problem in the energy field and it might well have applied also to nuclear technology had this not arisen as a by-product of weapons programmes.

UK energy use, past and present

460. Whether the world energy "gap" we have described will materialise and whether, if it does, the "gap" can be significantly filled by nuclear power, must be regarded as problematical. Certainly we should regard nuclear development on this scale as questionable not only environmentally but on economic and technical grounds. We now turn, however, to consider the energy position in the UK. For most of this century the principal source of energy in the UK has been coal, and it is only since the last war that oil, and since 1965, natural

gas have played any large part in supplying our needs. Figure 21 shows the energy usage per head in terms of primary energy and (since 1957) in terms of energy supplied, often in secondary forms, to final consumers. Until 1939, these two were almost the same as almost everybody bought coal and burnt it direct, often very inefficiently. The much increased material standard of living enjoyed nowadays has come about much more as a result of improvements in the efficiency of energy conversion and cleaner and cheaper fuels than because of increased use of primary energy. Two obvious examples are the replacement of coal-fired locomotives by diesel traction and of the domestic coal fire by gas or oil-fired central heating. As the figure shows, our net purchases of energy have hardly changed over the century, but our consumption of primary energy increased rapidly in the 1950s and 1960s with the import of increasingly cheap Middle East oil and more recently of very cheap North Sea gas.

FIGURE 21

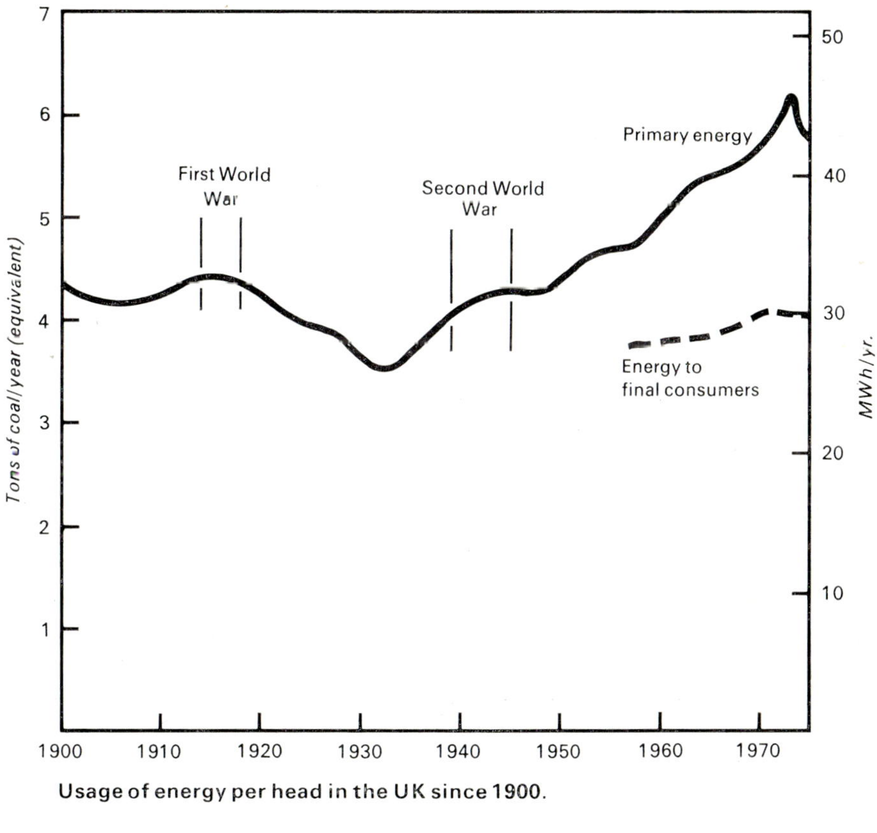

Usage of energy per head in the UK since 1900.

461. In 1975, Britain used somewhat less primary energy than in the "peak" year of 1973, as a result of the sharply increased prices which discouraged demand

and led to an industrial recession, and our total consumption was 2,374 TWh, made up as in the following Table:—

TABLE 12

Energy supplied to final consumers in the UK, 1975, TWh

User/Source	*Coal*	*Oil*	*Gas*	*Electricity*	*Total*
Industry	159	258	149	75	641
Domestic	127	42	173	89	431
Transport	—	356	—	3	359
Other	22	98	39	46	205
Sub-total	308	754	361	213*	1,636
Fuel industry losses	39	84	18	576	717
Stock changes	+68	—47	—	—	21
Total primary consumption (including that by electricity supply industry)	929	948	404	93†	2,374

*Approximately 5 TWh is used in the fuel industries. †Nuclear and hydro.

462. The Table gives a breakdown of energy supply to *final* consumers, but for some purposes the flow of primary energy is important, and this is shown in Figure 22. One of the largest users of primary energy is the electricity supply industry, and we give below the breakdown of its sources in 1975.

TABLE 13

The sources of energy used in the electricity supply industry in the UK*, 1975, TWh

Primary energy:		Uses of energy:	
Coal	514	Generation losses	534
Oil	157	Transmission losses	19
Gas	25	Industry use	18
Hydro†	12	(on works and for pumped storage)	
Nuclear	81	Sales	218
Total	789	Total	789

*Including purchases of nuclear electricity from the UKAEA and BNFL.

†The equivalent primary energy needed to generate hydro-electricity in steam plant.

The tables reveal that the electricity industry accounts for about one-third of all the primary energy used in the UK, though providing only about one eighth of the energy used by final consumers; the losses amount to nearly one quarter of our total primary use. We explained in paragraph 441 that such losses are inevitable when heat is converted into mechanical power.

463. About 40 per cent of the energy supply to final consumers is used for the servicing of buildings and it has been estimated that nearly 90 per cent of this

is used for low temperature heat for space heating and hot water. More than one-half of the domestic use of electricity shown in Table 12 is for this purpose, and so incurs heavy losses of primary energy.

FIGURE 22

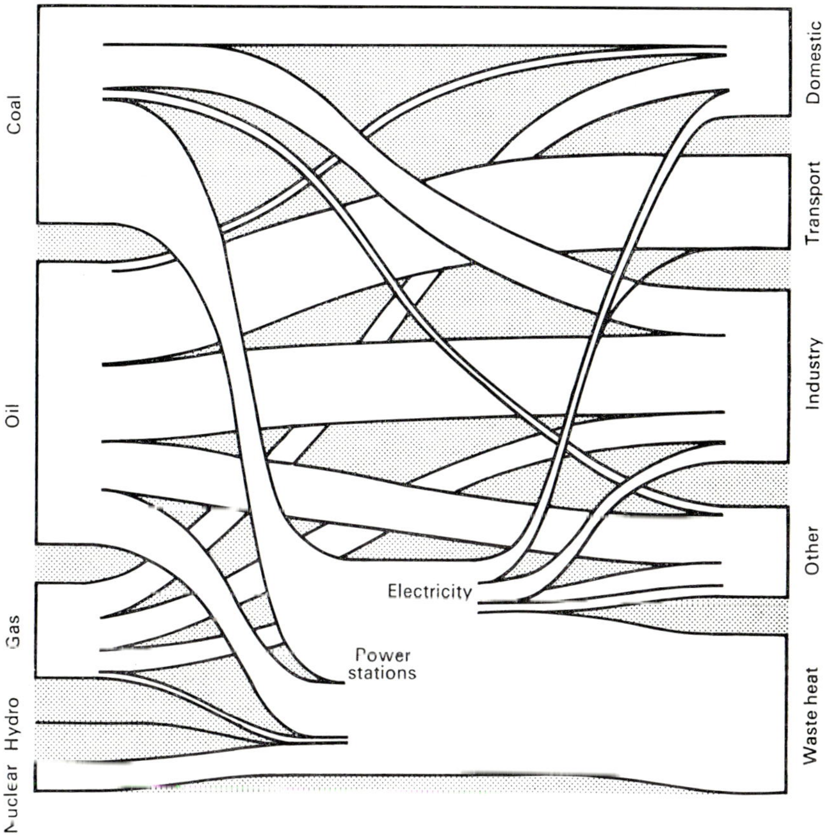

Energy flow in the UK in 1975
Vertical scale: 1 inch ≡ 750 TWh

464. In the transport sector, oil is virtually the only source of energy because of the dominant position occupied by road vehicles. This is unlikely to change unless new forms of electric storage battery are developed and introduced on a big scale. The efficiency of conversion from chemical energy to mechanical energy is generally quite low—of the order of 25 per cent. Thus the use of electric batteries would make little difference to the overall efficiency of energy use for transport, though the charging of vehicle batteries overnight would help the electricity supply industry to improve its load factor curve (see Figure 18).

465. Industrial use of energy, apart from the need for servicing buildings, representing about 133 TWh, is mainly for high temperature heat and mechanical power. There is some private generation of electricity within industry, about 17 TWh per year in addition to the 75 TWh purchased from the public supply. Many of these private generators also provide steam, process heat, or low-temperature heat for buildings in "total energy" or "combined heat and power" (CHP) installations.

Official energy strategy for the UK

466. The estimation of future energy supply and demand presents great difficulties because of the many uncertainties involved. Nevertheless because of the long lead-times needed for energy developments (such as the design and construction of nuclear reactors or the opening of new coal mines) those concerned with energy planning in government have to form a view on how the energy situation is likely to develop, and make appropriate plans. We invited the Department of Energy to let us have their estimate of how demand for energy might develop, and how this demand might be met. Since we were particularly interested in the nuclear component, we asked the AEA for their projection of this sector, subject to the likely constraints posed by uranium availability and plutonium formation in thermal reactors*.

467. The Department of Energy make a range of forecasts, based on different assumptions about the growth of the economy and on analysis of likely developments in demand in the various user sectors. Their recent discussion document on energy research and development requirements([62]) presents a range of possible options and introduces assumptions about the growth of the international economy and world energy prices in the future. The present view of the Department, as expressed in their evidence, is that it would be reasonable to assume that UK primary energy demand will continue to grow at a rate of about 2 per cent annually and that at least over the next 20 years the demand for electricity will grow rather faster, between say $3\frac{1}{2}$ per cent to $4\frac{1}{2}$ per cent although thereafter it may well slow down to a rate more in line with that for energy as a whole. The Department observed that much would depend on the extent to which electricity proved able to penetrate economically into uses such as transport and heating.

468. Within this overall demand for primary fuel, the Department calculated the amounts of coal, oil and natural gas that could reasonably be expected to be available, and found that there would be a "gap" of 370-750 TWh by 1990 and of up to 1850 TWh by 2000. These shortages (in primary energy) could be made up with from 20-45 GW of nuclear fission power stations in 1990 and up to two or three times as much in the year 2000. The AEA nuclear programme

*The AEA programme (104 GW in 2000) represents the maximum possible UK programme subject to these constraints, but we were surprised to discover that it was considerably smaller than that originally advanced in their evidence (149 GW in 2000) and much smaller than that calculated as desirable in a companion document (210–285GW in 2000).

(see Table 14 below) would provide for 30 GW by 1990 and 104 GW by 2000: it thus falls within the Department's estimates of requirements.

TABLE 14

AEA—projected programme of electrical generating capacity installed at end of year indicated, GW

Year	*Thermal*	*Fast*	*Total nuclear*	*Fossil*	*Total*
1975	5	—	5	66	71
1980	11	—	11	77	88
1985	14	1	15	83	98
1990	26	4	30	97	127
1995	45	19	64	96	160
2000	71	33	104	94	198
2010	104	89	193	101	294
2020	104	200	304	107	411
2030	56	370	426	129	555

469. The timescale of this projection is sufficiently long that it will cover the period when oil and gas from the North Sea are likely to be in very short supply. The Department made no specific allowance for the advent of nuclear fusion or for the development of electrical capacity based on solar, wind, wave or tidal power. In their oral evidence to us they expressed the view that, were electricity to be needed in the amounts foreseen, these sources, even if successfully developed, could make only a marginal contribution and would not significantly alter the magnitude of the required nuclear power programme. If fusion were successfully developed so that commercial stations could be ordered in, say, 2010 for commissioning in 2020, the effect would be to reduce the number of fast reactors thereafter, but it would make no essential difference to the official strategy.

470. The figures in Table 14 give electrical generating capacity which must of course be sufficient to cover peaks in demand. In order to assess the amount of electricity that would actually be generated it is necessary to make assumptions about variations in demand and about the extent to which the rated output of generating plant can be achieved in practice. The AEA supplied estimates of the amount of electricity that would be generated under the programme described in the Table and we have used the estimates in deriving the relevant curves in Figure 23, in which we have identified the main sources of energy supply up to 2025 that correspond to the official projection we described above.

471. We comment in passing on another aspect of this nuclear programme. It has been suggested in evidence to us that a rapid growth in nuclear power, such as this programme undoubtedly represents, might actually lead to a negative net energy output; that is, that over the growth period more energy would be used in constructing the nuclear plant than would be obtained as electrical output. We have made some calculations to satisfy ourselves on this point. It turns out that, on the assumption that 10 TWh of thermal energy are needed to construct 1 GW of electrical plant([63]), the AEA's programme would produce a net positive output at all times, and that the percentage of the apparent output

FIGURE 23

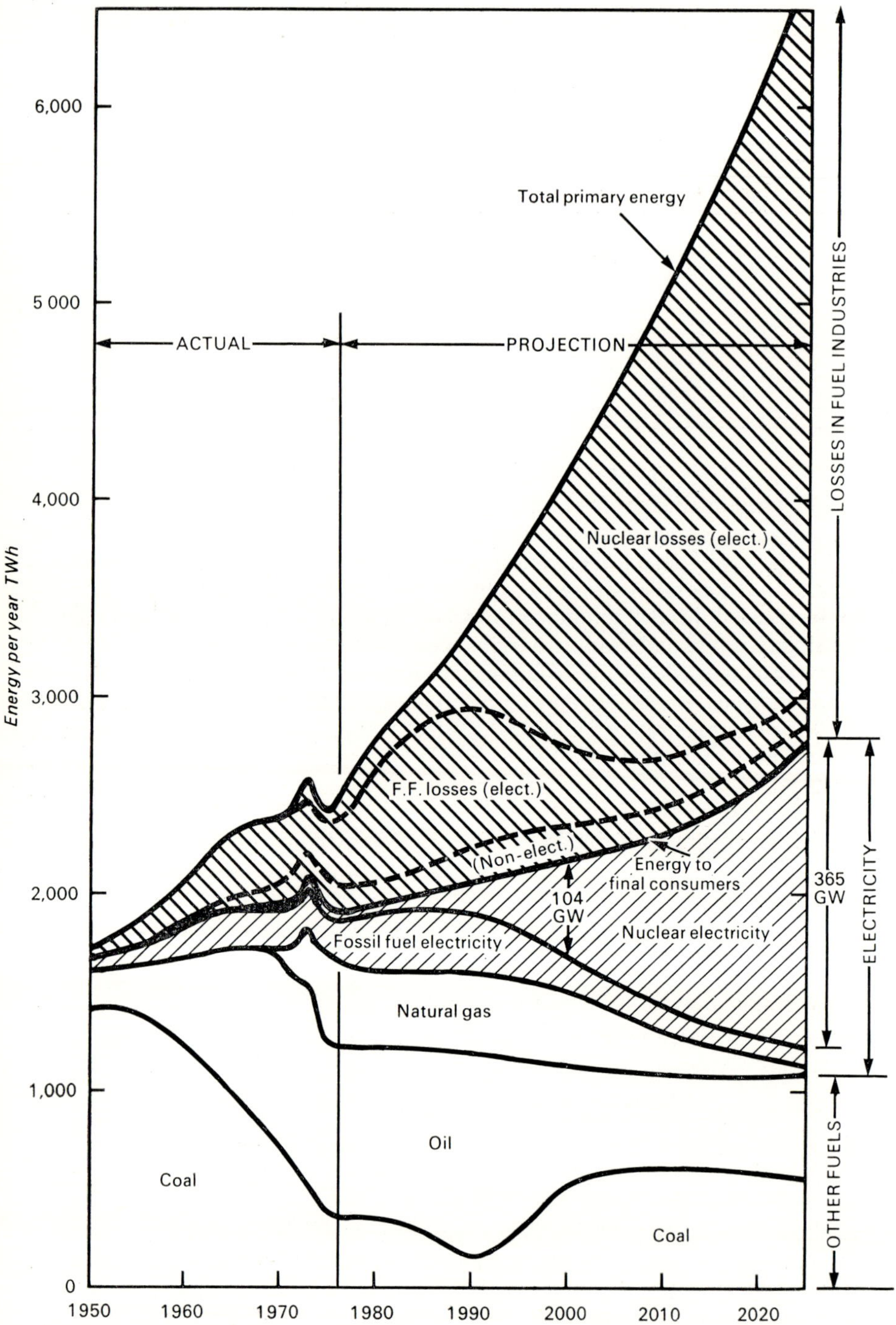

The provision of energy for the UK, projected to 2025, on the 'official strategy'.

used in plant construction would drop from about 20 per cent in the 1980s to 10 per cent in the next century. There may be other grounds for querying the programme, but this is not one.

472. The non-nuclear fuels shown in Figure 23 are coal, oil and gas, and it appears that all of them can be found from our indigenous resources; the amounts are broadly consistent with the estimates given in the Department of Energy's evidence. We have assumed in making the projections that the National Coal Board's "Plan for coal" is fully implemented for the next ten years, and that production reaches 150 million tonnes (1100 TWh) in 1990, but that it gradually declines as a result of the big investment in nuclear power to about 110 million tonnes (814 TWh) in 2025. In the later years it will be necessary to divert coal from electricity generation to the production of synthetic natural gas and to liquid fuels suitable for transport applications.

473. Oil and gas are expected to make a large contribution in the early years of the period. We have taken the amounts ultimately available from the North Sea to be respectively 4,000 million tonnes (48,000 TWh) and 60 billion cubic feet (17,600 TWh), so that for there to be significant amounts remaining in the next century, when they will be extremely valuable, a vigorous policy to control the depletion rate will be needed in the early years of exploitation. Oil will be needed not only as a fuel, but also as a chemical feedstock, and we have allowed for an increase in this requirement in calculating the amount needed to be refined, and hence the total demand.

Discussion on official energy strategy

474. The possibility that energy demand and the supplies available from fossil fuels may develop as portrayed in Figure 23 is taken as indicating the probable necessity for so large a nuclear programme in the future. There is, of course, no commitment to such a programme; nuclear plant would be installed gradually in the light of changing assessments of likely electricity demand. At the present time there is a substantial surplus of electrical generating capacity and it might well appear that anxiety about a predominantly nuclear energy future is premature. Nevertheless, because of the long lead-times involved in energy planning, the fact that it is envisaged that supply and demand might develop along the lines indicated by this projection will affect present plans and priorities, and might do so in such a way as gradually to foreclose other options that would have been available had they been exercised in time.

475. It is assumed in the official strategy that electricity will steadily increase its penetration of the energy market, to the extent that by 2025 electricity would account for about 60 per cent of energy supply to consumers. For the reasons given below we should regard this as undesirable on environmental grounds, but we note first that the assumption is highly questionable in that it appears to ignore the effect of increasing price on demand. Experience since the early 1950s shows that even modest rises in prices of energy relative to other commodities may be expected to have a depressing effect on demand; this effect is illustrated in the domestic sector by Figure 24. The recent very large

increases in the price of electricity have resulted in a decline in consumption, particularly for heating where even off-peak electricity at 1p per unit (which itself is heavily subsidised) cannot compete with gas at less than half the price.* This trend is most unlikely to be reversed unless at some time in the future electricity were once again to be comparable in price with gas, when the lower installation costs associated with electric heating might make it attractive.

476. The technical cost of natural gas from the North Sea is low, however, and unless it is severely taxed it may be expected to be very competitive for several decades. Even when it has to be replaced with SNG made from coal, gas is likely to be substantially cheaper than nuclear electricity for heating purposes. The very high energy requirement for the construction of nuclear stations means that fossil fuel prices will be reflected in the price of nuclear electricity, and the price of uranium may also rise. Thus, even a large increase in the price of fossil fuels could probably be absorbed by an SNG industry before the differential would be eliminated. We see no basis, therefore, for assuming the early penetration by electricity of this important sector of the energy market. Moreover, as noted above, sharp price increases would depress demand so making it questionable whether the growth rate implied in the strategy would be sustained.

477. We think that there are also grounds for serious doubt about whether the nation could sustain a nuclear programme on the scale indicated in Table 14. We have roughly assessed the investment implied by this programme in relation to GNP. For this purpose we have assumed an annual growth in GNP of 3·3 per cent for the ten years from 1975 and a growth rate thereafter of 2·7 per cent (these two figures correspond to the upper and lower growth rates assumed by the Department of Energy for their projections, and the latter figure is characteristic of the UK economy over the last 20 years). We have taken the capital requirement for nuclear stations as £750/kW in 1975 £; this figure is based on the investment approval figure given in Parliament for the SGHWR stations, plus interest during construction (at 10 per cent) for 3 years, the costs of associated fuel cycle facilities (10 per cent) and of transmission and distribution (40 per cent, which may be an underestimate as it has been between 70 and 115 per cent during the last 15 years). On this basis the investment required for electricity generation in 1995 and beyond would amount to 3·4 per cent of GNP compared with an average level in the past 20 years of 1·4 per cent. We are very doubtful whether this would be compatible with the investment in other sectors that would be needed to create the demand for the electricity generated.

478. Another reason for questioning whether such a nuclear programme would be practicable arises in connection with the siting of the many new power stations that would be required. Since it is envisaged that these stations would be used only for electricity production, they would generally have to be sited on the coast to obtain adequate supplies of cooling water. We have considerable doubts about whether enough coastal sites could be found for the 426 GW of

*In 1965, off-peak electricity was actually cheaper than the mean price for gas to domestic consumers, and sales of night storage heaters consequently rose very fast.

FIGURE 24

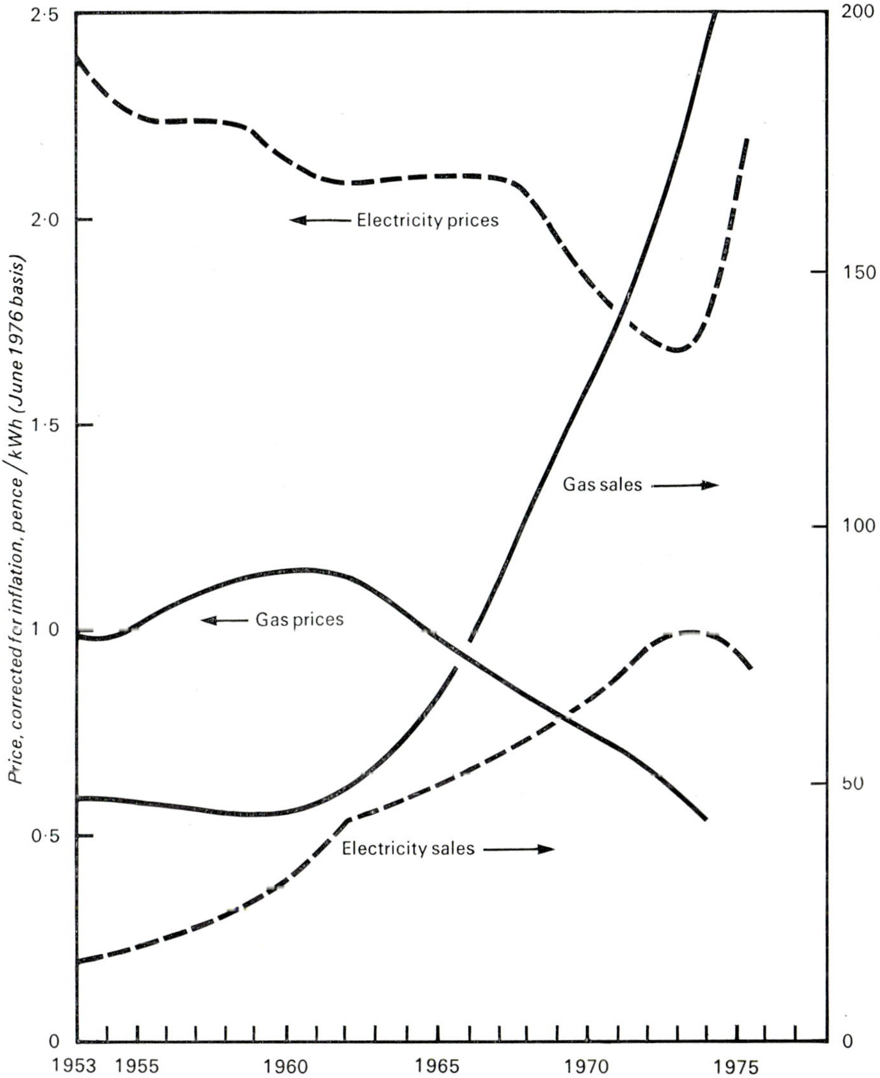

The effect of price changes on sales of gas and electricity to the domestic market in England and Wales since 1953

nuclear power (about one hundred times our present nuclear capacity) that the AEA foresee in 2030, having in mind both the stringent siting criteria and the objections that there would be on amenity grounds to many potential sites.

479. For these reasons we regard the approach to future energy supplies that forms the basis for official strategy as unconvincing. Still, we must assume that such a future might materialise, and consider its environmental implications. We have explained at length in this Report our anxiety about the problems and risks that would be created by such great dependence on nuclear fission power. There are, however, other and substantial environmental objections to the strategy we have described. The first of these is that such a high-electric future would be exceedingly wasteful in energy terms, as is apparent from Figure 23; in 2025, for example, about one-half of the primary energy converted would be wasted, largely in the form of rejected low-temperature heat at power stations. From the environmental viewpoint (as well as others) such waste is to be deplored, because it almost invariably means more pollution.

480. One aspect of this is the effect on the environment of the heat arising from energy conversion. Energy in all its forms eventually appears as heat in the environment, and as energy use increased this heating could have noticeable effects on weather patterns and climate. It has been suggested([64]) that such effects would occur at levels of energy use a few times that of the present in the UK (where it already exceeds 1 per cent of the solar flux), though the Department of Energy and the CEGB (in their written evidence to us) dismissed such fears as groundless.

481. The Commission reviewed the effects of heat release, as well as those of increased carbon dioxide and particulate formation that would be associated with the burning of fossil fuels, in their First Report (paragraph 116). At the time it appeared that they were likely to be insignificant in terms of world climate. However evidence we have received from the Meteorological Office suggests that there might well be regional effects, for example in Western Europe or the USA, if energy use were to increase above present levels by, say, an order of magnitude. Oceanic heat source anomalies are known to cause year to year variations in atmospheric circulation. These are of the order of 10^8MW, and the Meteorological Office consider that significant effects would arise with a regional energy use of only a fraction of 10^8MW. Gross primary energy usage both in Western Europe and the US is at present about 2 per cent of this amount, and would reach 10 per cent in about 50 years, if it were to increase at 3 per cent per year. This suggests that regional climatic effects should not be ignored when considering energy policy for this time scale. The Meteorological Office acknowledged that significant effects might occur even earlier on a national or local scale.

482. They also stressed the great uncertainties attaching to any judgment of these effects, and the need for much more study before confidence could be placed on them. Nevertheless, it would appear to us to be prudent for some account to be taken in official energy strategies of the risk that a very large energy release may have adverse climatic effects. The risk provides one reason

for seeking to reduce the waste of energy referred to above. Our concern is not with the possibility of there being small changes in mean temperature, which seem to occur naturally over periods of decades and centuries, but with major changes in temperature and circulation, and in consequential temperature and rainfall patterns, which could have serious implications for agriculture and the natural world.

483. In paragraph 478 we mentioned the siting problems that were likely to arise for the nuclear stations that would be required by the official strategy. Such a programme would have substantial adverse effects on amenity whether the stations were sited on the coast or inland, when the huge cooling towers that would be needed would dominate the landscape in many areas of the country. Despite careful landscaping, as at Didcot, these structures are inevitably very intrusive, and besides, they occupy large areas of valuable land. Further major extensions of the 400 kV grid would be required, leading to very many more transmission pylons and overhead wires. Few areas of the countryside in Britain would be free from the outward and visible signs of electricity generation and distribution.

484. There thus appear to us to be very considerable environmental objections to the high-nuclear, high-electric, energy future that is foreseen in the official strategy. We thought that it would assist public understanding of the issues if we were to give some thought to the possibility, and the environmental implications, of an alternative strategy that would provide the same energy to final consumers. It would be an aim of this strategy to reduce waste by seeking to match the kind of energy supply to user requirements, and to reduce future dependence on nuclear power.

An alternative energy strategy

485. The strategy is illustrated in Figure 25. As a necessary starting point, we have had to attempt an estimate of the extent of the future need for electricity as a proportion of the energy supply to final consumers. For this purpose we have considered each of the demand sectors (industry, domestic, transport and "other users") and have made simple, and necessarily tentative, assumptions about the way in which electricity demand might be expected to change. It will be seen from Figure 25 that the outcome is a growth in market penetration by electricity from 13 per cent in 1975 to 19 per cent in 2025, as compared with 60 per cent for the official strategy.

486. Most of this electricity would be provided by nuclear stations, but we have assumed that a substantial proportion would eventually be generated from natural sources, particularly waves. This implies a judgment that the technology will be available on the time scale indicated, but we think that the technical difficulties, although not to be minimised, need not necessarily be more daunting than those associated with the successful development of a demonstrably safe fast reactor and its fuel cycle.

487. There may also be environmental problems with these new sources of electrical power. Aerogenerators ("windmills") would need to be prominently sited, and if installed on a large scale would damage amenity, though perhaps no more so than the electricity pylons that already exist. Wave power devices seem unlikely to have a significant environmental effect in themselves, but they could constitute a danger to navigation unless they are sited off shipping routes, which seems possible. A more severe impact is likely to arise from the need to transmit the electricity from the Hebridean islands down through the Scottish Highlands to the load centres further south.* Tidal barrages can be expected to have a fairly dramatic impact upon the estuary in which they are situated, but each one would have its own problems. Some of the effects could be beneficial, as there will be new facilities, e.g. for sailing and other water sports—but these may in turn have a disturbing effect upon wild life in the area. In the Severn estuary, which is the one most likely to be exploited, careful studies would be needed to predict the effect of a barrage on pollution, and on land drainage.

488. The difficulties in matching supply and demand in strategies that do not involve a major extension of electricity are in the provision of adequate amounts of heat and portable fuels, which are at present provided largely by oil. We have assumed substantially the same use of oil and natural gas as in the official strategy. In the long term the sun would be an important source of supply for heating requirements; it has been estimated([65]) that solar power could provide some 260 TWh by 2020. Over much of the period we are concerned with, however, there would need to be increased reliance on coal. Britain is fortunate in having abundant reserves, and an industry that is well organised to exploit them. The National Coal Board gave evidence to us and told us of significant advances in both extraction and utilisation technology. They said that it would be reasonable, given the necessary capital investment, to expect coal production to rise over the next 20–30 years to 200 million tonnes per year and to be maintained at this level for many decades.

489. Coal mining on this scale (about half as much again as at present) would clearly create a number of environmental conflicts, which are exemplified by the consequences of the Selby mine as revealed at the recent Public Inquiry. To reach an annual output of 200 million tonnes, about five or six new mines the size of Selby would be needed, or their equivalent in terms of expansion of output from existing collieries, in addition to what would be needed to replace falling output from worked-out seams. Mining produces subsidence and often creates unsightly—or even dangerous—spoil heaps which require landscaping. But these effects can be mitigated with modern mining techniques. For example, we understand that it is possible for the minestone produced to be returned to the mine and that this technique is used in Poland to prevent subsidence in urban areas. Minestone could also be used in much larger quanti-

*An alternative would be to generate hydrogen and transmit this by pipeline in order to hydrogenate coal for the production of SNG and liquid fuels. This may in fact prove more satisfactory, as it would provide a degree of storage to circumvent the problem of accommodating calm periods in winter.

FIGURE 25

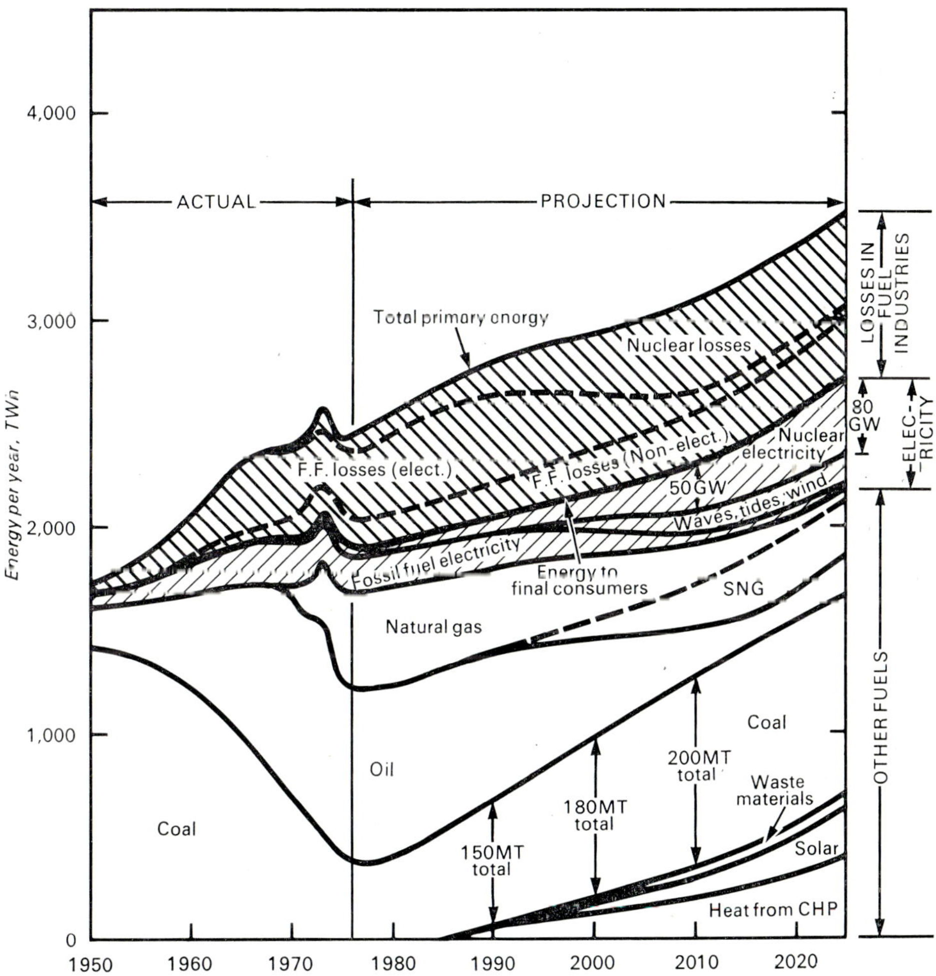

The provision of energy for the UK, projected to 2025, on the 'alternative strategy'.

ties as an alternative to natural aggregates. In the Selby mine, very little minestone will be produced as the coal seam is thick enough to permit machinery to operate entirely within the coal.

490. The combustion of coal has traditionally given rise to much pollution, but we do not think that this need be so in the future if some of the current developments, including in particular fluidised-bed combustion,* are brought into widespread commercial use. The pollution from coal conversion plants is more likely to be in the form of dirty water, and it may be expensive to provide treatment. However, since such plant is now being ordered for areas of the USA that are already short of water, with a requirement that the water should be re-used within the plant after treatment, we do not think that this problem should prove insuperable. The plant would not be needed in any quantity here before about 2000, so there should be adequate time for its development. By-products of such plant could well include organic chemicals which would serve as feedstocks for the chemical industry. There is a loss of energy involved in converting coal to SNG or liquid fuels, but overall the process is more efficient than burning it in power stations to generate electricity (unless these are of the CHP type).

491. The other source of heat which we have assumed in the development of the alternative strategy is district heating by means of piped hot water, mainly from stations that also generate electricity as a by-product. We are aware that this process has been generally found uneconomic in this country, although it has been adopted in many continental countries and there are connected heat loads of 9 GW in Sweden, 14 GW in France and 20 GW in West Germany. The Department of Energy have set up a committee under their Chief Scientist to examine the scope for such systems. In the discussion document referred to earlier, it was envisaged that district heating from CHP plant might provide up to one quarter of the heat needed in buildings by 2025, that is some 240 TWh, and that there would be perhaps half of this amount used in industry at that time. Other writers have suggested more ambitious market penetrations. At present it is doubtful if district heating schemes would be competitive with natural gas for heating where this is available, but the position will change when natural gas has to be supplemented to a large extent with SNG from coal, and when the need to conserve fuel will be more apparent. Here, as elsewhere in matters of energy supply, present economic reasoning may support policies that would prove unwise in the long run. The future benefits of having district heating do not at present look economically attractive when discounted at 10 per cent per annum, but this should necessarily be used to prevent a start being made on CHP projects in the more obviously suitable areas such as city centres. The same argument could equally well be used against the FBR programme, whose economic benefits will be most apparent when uranium is very expensive.

*Fluidised-bed combustion is a process in which the solid particles of fuel and ash are maintained in a fluid condition by blowing air through them from the base of the furnace. The combustion process is very efficient and many of the impurities in the coal (notably the sulphur) are retained in the ash, so that the process causes much less pollution than from normal combustion.

Discussion on alternative strategy

492. We have sketched the outline of an alternative approach to energy supply and we are fully aware of the speculative assumptions on which it is based. We certainly do not present it as a seriously studied option, though we have sought informatton on a number of features involved in the strategy (for example, on the potential and practicability of wave power and of CHP systems) from people with expert knowledge. Our purpose has been simply to form a view on whether a major future expansion of nuclear power, involving the widespread use of fast reactors, must be seen as inevitable, given the projected increase in energy demand. The conclusion we have reached is that the case for such expansion is by no means so clear-cut as has been represented to us by the official bodies reponsible for energy policy.

493. Our analysis has taken no specific account of energy conservation measures, apart from the particular aspect of the use of CHP systems. This is also true for the official strategy, and indeed conservation was not referred to in the Department of Energy's formal evidence to us. A number of studies have been made of the potential for energy saving through conservation. Thus, the Select Committee on Science and Technology suggested(66) that a saving of 15 per cent could be made without major changes. Sir William Hawthorne, the Chairman of the Department's Advisory Council on Energy Conservation, said in his evidence to us that savings of 25 per cent over 25 years might be possible if a major effort was made; this would *halve* the Department's projected rise of 2 per cent per year in primary energy. We have referred earlier to our objection, on environmental grounds, to the waste of energy and we hope that the Council will be influential in reducing it. Any reduction in demand achieved through conservation would increase the prospect that energy requirements could be met by a strategy on the lines we have proposed.

494. The alternative strategy is designed to provide the same energy to final consumers as the official one, the latter being based on an annual growth rate of 2 per cent in primary energy demand. Depending on the assumptions made, many other projections are possible of the way in which energy demand and supply may develop. A number of options are considered in the Department of Energy's discussion paper(62) and we welcome the fact that one of these is a non-nuclear option under which existing reactors would not be replaced at the end of their useful lives. (It is assumed, however, that the rest of the world adopts similar policies.) On this projection, growth of the UK economy would be a little less than in the recent past (2·5 per cent instead of 2·7 per cent) until 2000, and drop below 1·5 per cent by 2020. Rigorous conservation measures would be needed and there would of course be a lower growth rate in energy supply to consumers than in the projections we have considered. However, if allowance is also made for the probable future decline in the rate of population growth, the rate of growth in energy supply per head would be little different from what has prevailed over the recent past.

495. We are not advocating a non-nuclear future, though we think it is highly desirable that there should be the fullest discussion of the implications

of this and other energy options. Our charge is the protection of the environment and we would favour an energy strategy that offers the prospect of least environmental harm. We recognise, however, that other conflicting requirements have to be satisfied. One of these is security of supply and the desire to insure against the uncertainties of relying on supplies from abroad. Another is cost; thus a decision not to use fast reactors might mean that at some time in the future electricity would be more expensive. If electricity prices were to rise in real terms by, say, 5 per cent by 2025 as a result, this might be accepted as a small price to pay to avoid the risks posed by fast reactors and plutonium. However, if prices were to double by 2010, then there would certainly be more argument.

496. Our anxiety about the hazards of an economy based on plutonium leads us to the view that fast reactors should be introduced only if they are demonstrably essential. Judgment on whether their use is essential would need to take account of factors such as those referred to above, which we cannot authoritatively assess. These factors apart, however, our alternative strategy suggests that fast reactors might be avoided at least over the fifty-year period covered. The contribution we have assumed from nuclear power could be supplied by thermal reactors, for the total uranium requirement would amount to only about one-half that of the thermal reactors needed under the AEA programme.

497. We noted previously that nuclear expansion is a very long way off and that it may appear too soon to worry about these issues, but that thinking on energy strategy affects present plans and priorities. We believe that there have been significant developments in Department of Energy thinking since its original submission of evidence to us, but the impression we formed from that evidence and from our discussion with officials was that the Department's policy had been essentially passive. The underlying attitude appeared to be that thermal reactors were available, that fast breeder reactors could in time be made available, that our future supply position was therefore basically secure and that developments could be left to take their natural course. Such an approach would not be compatible with our judgment of the strength of the case for seeking to avoid a large future commitment to nuclear power.

CHAPTER X

NUCLEAR POWER AND PUBLIC POLICY

Introduction

498. In previous chapters we have considered the environmental implications of nuclear power, both actual and potential. We now draw the various considerations together in discussing the future of nuclear power and in giving our views on the policy that should be adopted towards its development. As we said at the start of this Report, we are conscious in doing so of touching on matters that are at the fringe of, if not beyond, our strict terms of reference as a Commission concerned with environmental pollution. This appeared to us to be unavoidable, however, if we were to present a complete account, and to provide a foundation for the wider discussion of these issues that we believe to be essential.

499. It is important that the discussion should be well-informed and as far as possible dispassionate. It seems that nuclear power has in some ways become the whipping-boy for technological development as a whole. Many thoughtful people believe there is a progressive deterioration of the environment which they associate with the spread of technology (we have in mind here not pollution alone, in which there have been many local improvements, but the total impact on society) but feel unable to make a stand against this process because each individual step is small and can probably be justified as bringing benefits at minimal environmental costs. Nuclear power provides a dramatic focus for opposition in some countries to technological development and we have no doubt that some who attack it are primarily motivated by antipathy to the basic nature of industrial society, and see in nuclear power an opportunity to attack that society where it seems likely to be most vulnerable, in energy supply. In the USA in particular we see a situation in which the debate between the nuclear industry and the environmental movement has become increasingly vociferous. The environmentalist tends to see those in the industry as being so committed to furtherance of their technology as to be wilfully blind to its dangers to the world. Those within the industry, many no doubt sustained by the thought that they are thereby making an essential contribution to the well-being of mankind, tend to see environmentalists as people opposed to all technology who are prepared to denigrate their work on the basis of drummed-up and nebulous fears of future catastrophes. During our study we have had many contacts with the various sides of the nuclear industry in this country and with its environmental critics. The arguments of both deserve to be heard with greater mutual understanding.

500. We are acutely aware that concern about the large-scale use of nuclear power, and the concomitant threat of plutonium in a civilian context, may appear misplaced in relation to what many would see as a much greater threat, that growing arsenals of nuclear weapons may one day lead to nuclear war.

The implications of civil nuclear development are qualitatively different, however, and are not to be seen as a mere increment in the nuclear threat. The connection between civil and military uses of nuclear power is obvious; there is the possibility that great expansion of civil use would make more difficult the eventual acceptance and introduction of measures to secure nuclear disarmanent.

501. We begin our discussion on the future of nuclear power by recapitulating the main hazards of its development and the salient features of future energy supply.

The hazards of nuclear power development

502. There is first the question of reactor safety. We have seen that the consequences of a large release of radioactivity could be extremely serious (paragraphs 265–268), and concern on this point is not entirely assuaged by the fact that other technologies may present comparable, or even greater, hazards. We are confident that the risk of accident leading to a serious release of radioactivity is at present very small, having in mind both the existing scale of nuclear provision and the attention devoted to safety in reactor design and operation. The risk will certainly increase with the number of installations, however, and a rigorous commitment to safety must be maintained. There is the question whether this commitment might become dulled; it is a common enough experience in other fields that familiarity, and continuing success in an enterprise, lead to failures in preserving standards and adequate organisation. The Flixborough disaster supplies an example of this. There is the question of the potentially greater hazards of the fast reactor (paragraphs 297–304) and whether, in our anxiety to grasp the benefits it offers, we may be tempted to accept a qualitatively different increment in risk. There can be little doubt that as the number of reactors in use throughout the world increases, there will in time be a major accident. Such an accident, even if it occurred abroad, would certainly cause anxiety in this and other countries and might require reactors to be shut-down while safety aspects were reassessed. This alone suggests that it would be prudent to aim for diversity in energy supplies.

503. The risk of a serious release of radioactivity from a reactor or other nuclear installations would be substantially increased by act of war or sabotage. Nuclear installations could well be prime targets in time of war. A policy of concentrating nuclear facilities on one site, such as might be adopted to reduce the problems of safeguarding dangerous nuclear materials which would otherwise have to be transported from one facility to another, would make the site more vulnerable and more attractive to attack. There is here a difficult question of balance.

504. We have described in Chapter VIII the problems associated with the management and disposal of highly radioactive wastes arising from the nuclear fuel cycle. Such wastes will remain active over immense time scales, and unless continuously isolated will present dangers to our remote descendants long after nuclear fission technology has ceased to be used as a source of energy. We

believe that a quite inadequate effort has been devoted to the problems of long-term waste management, and that there should be no substantial expansion of nuclear power until the feasibility of a method of safe disposal of high level wastes for the indefinite future has been established beyond reasonable doubt. There are promising ideas for the disposal of these wastes but it may take 10 to 20 years to establish their feasibility.

505. Lastly, there is the problem, discussed in Chapter VII, that is in many ways the most worrying and the most difficult to assess, that is, the risks arising from the production and use of plutonium in large quantities, which would follow from the widespread introduction of fast breeder reactors. This proliferation would make it easier for terrorist or criminal groups to obtain the element and use it in acts against society. Knowledge of plutonium and the ability to use it for nefarious purposes will inevitably be disseminated as nuclear power spreads. There is no lack of demonstration in the world at present of the audacity, determination and ruthlessness of terrorist organisations. Unless we are prepared to assume that terrorism is no more than a transient phenomenon, or that terrorist groups would shrink from using the immense threat of plutonium to achieve their ends, then the future risk of such action exists and must be considered.

506. Another aspect of the problem is the possible effect of such threats on society. The security measures that might become necessary to protect society could seriously affect personal liberties. The need for such measures would be affected by increasing tensions between nations. Indeed, the future risks posed by plutonium constitute a world problem that would not be solved by unilateral action by the UK, though the action we take in response to our assessment of these risks could have a substantial impact on world opinion. We emphasise again that our concern here is not with the position at present, or even in the next decade, but with what it might become within the next fifty years. In speculating on developments on such a time scale no one has a prerogative of vision. It appears to us, however, that the dangers of the creation of plutonium in large quantities in conditions of increasing world unrest are genuine and serious.

507. For this reason we think it remarkable that none of the official documents we have seen during our study convey any unease on this score. The management and safeguarding of plutonium are regarded as just another problem arising from nuclear development, and as one which can certainly be solved given suitable control arrangements. Nowhere is there any suggestion of apprehension about the possible long-term dangers to the fabric and freedom of our society. Our consideration of these matters, however, has led us to the view that we should not rely for energy supply on a process that produces such a hazardous substance as plutonium unless there is no reasonable alternative.

508. We have considered whether this country should seek to abandon nuclear fission altogether, but even if it could be confidently supposed that this could be done without risk of unacceptable restrictions on energy supply in the future, we should not think that such a strategy was wise or justified. We do not think

that continued operation of the existing Magnox reactors, or the eventual operation of the thermal reactors that are now being installed or considered, can be thought to constitute an unacceptable hazard when judged against the benefits, provided that existing standards are maintained, as we have every reason to believe they will be. The risks we have described exist but they must be balanced against the desirability of diversity in sources of electricity. There are already substantial quantities of radioactive waste requiring secure disposal, but we are sufficiently hopeful that an acceptable solution will be found to this problem that we do not advocate halting the processes giving rise to the waste. Indeed, to do so might remove the sense of urgency that we believe to be required in seeking a solution. Finally, if as we hope fusion power is eventually developed, it will require the technical expertise and experience obtained through fission technology for its successful introduction and operation. The abandonment of fission power would certainly not advance the advent of fusion and would probably delay it.

Aspects of future energy supply

509. Assessment of whether or not there are reasonable alternatives to dependence on fission power in the future leads to consideration of energy forecasts. In the previous chapter we described the programme of nuclear expansion envisaged by the Department of Energy and the AEA. Their estimates indicate a widening gap between energy demand in the future and the contribution that can be expected from fossil fuels. In their view nuclear power was the only proved technology that could fill this gap. Even allowing for the development of other sources of energy and the effects these might have on the scale of nuclear provision, the Department considered that a substantial fission programme would be needed; and that because of uncertainties about the future availability of uranium supplies it was essential to keep open the option of using FBRs, and hence to proceed with their development.

510. We accept the need to retain a nuclear option but, as we explained in Chapter IX, we are concerned about the projected nuclear expansion not only because of the specifically nuclear risks involved but on general environmental grounds. We felt bound to seek some understanding of whether such expansion was inevitable or whether alternative strategies might be available for meeting future energy needs. Our consideration of this matter has been necessarily limited and tentative, and we cannot be sure that the approach we have outlined would be feasible. We fully realise that factors we have been unable to consider may crucially affect the issue. One factor of this kind, which is environmental in nature, is the build-up of carbon dioxide in the atmosphere from the combustion of fossil fuels. We referred to this problem in paragraph 192; it could yet provide a powerful argument for nuclear development. We regard the strategy we have described as doing no more than suggest that the case for a nuclear programme involving the installation of many FBRs is not yet proven. Whatever the merits of this strategy it is important that alternatives should be considered and this has been recognised by the Department of Energy which is now looking at a wide range of energy options. Even if an alternative strategy could not give the same assurance that future energy demand could be satisfied, there is

the possibility that the nation would prefer to accept some risk in order to avoid what might be seen as the greater risk of a much expanded nuclear future.

511. Clearly in the long run other sources of energy will need to be developed, some now probably unsuspected. We have referred to one of these, namely nuclear fusion. Compared with fission, fusion would involve much less hazard from radiation. Fusion power has its critics([67]); it is likely to be very capital intensive and to become available only in very large units applicable only to the centralised generation of electricity. We mention here that we attach importance to the possibility of developing fusion reactors of moderate size which would be suitable for urban siting and for district heating purposes. To this end we think that the possibilities of laser fusion should be explored. Our basic concern is that a major commitment to fission power and the plutonium economy should be postponed as long as possible, in the hope that it might be avoided altogether, by gaining the maximum time for the development of alternative approaches which will not involve its grave potential implications for mankind.

512. It may well be said that our concern on these points is premature; that there is, after all, no firm commitment to nuclear provision on the scale indicated in the AEA projections, and that other energy sources and ways of using energy more efficiently are already being actively examined. We should not be satisfied with this response. The important thing is the attitude towards alternative approaches and the resources devoted to assessing their potential and promoting their development. The basic belief of the Department of Energy and the AEA is that nuclear fission using the fast breeder reactor is the only real option for meeting our future energy needs. We fear that on this premise there may be a gradual, step by step progression to overriding dependence on nuclear power through tacit acceptance of its inevitability, and a gradual foreclosing of other options that might have been available had they been exercised in time. We recommend that it should be an aim of policy to lessen our future dependence on fission power to the extent that this would command public acceptance in the light of a full understanding of the implications and of the issues involved. We return to this point in paragraph 521.

513. In our view such a policy would certainly call for increased support for the development of alternative approaches and energy sources. A programme of research on wave power was recently announced which will cost about £1 million over two years. A sum of less than £0·5 million per year is currently being spent on other sources, including solar energy.* In contrast, annual expenditure by the AEA on nuclear fission research and development amounts to about £80 million at present. The proportional expenditure on alternative sources is substantially smaller in the UK than in most other industrialised countries. The Department of Energy are examining, we think rather belatedly, the possibilities afforded by combined heat and power systems

*The Department of Energy announced on 22 July 1976 that an additional sum of £840,000 was to be allocated for a three-year programme of research into the possibilities for geothermal heat.

which are already in substantial use in some other countries. We think that increased resources need to be devoted to energy conservation. The Department have established the Advisory Council on Energy Conservation and we greatly welcome this, for we are not convinced that conservation has been pursued with sufficient vigour. Among the alternative sources we include fusion, and we are agreed that on environmental grounds, there should be substantially more effort directed to the development of fusion power. The UK is currently spending much less on fusion development than some other comparable developed countries such as West Germany and Japan.

514. We have given some thought to whether the present organisational arrangements are conducive to achieving an appropriate balance in the use of resources for energy research and development in the different fields. In this context we have noted the conversion of the former AEC in the USA into the Energy Research and Development Administration whose annual budget covers support for both nuclear and non-nuclear energy projects and which we believe has led to larger and better co-ordinated projects in the non-nuclear field. We do not suggest that an identical body would be appropriate in the UK where the structure of organisations involved in energy matters is very different. In the USA, for example, the coal industry is fragmented, in contrast to the position in this country where research and development in this area is conducted by the National Coal Board. However, we see advantage in extending the remit of the AEA so that it would become responsible for energy research and development in a wider field. We have in mind that the Authority would assume responsibility for work on renewable energy sources. We would expect much of this work to be done under contract from the Authority to research groups in universities and other establishments, the programme being controlled and co-ordinated by the AEA. The AEA would need to build up relevant expertise and we foresee that, at a later stage in development, some of its present scientific and technical resources would be diverted to work in these other fields. We recommend that the Government should consider this extension of the AEA's functions and correspondingly of its title.

515. In the end the allocation of resources for energy development must depend on energy policy. We are conscious of the immensely difficult issues involved in framing a coherent policy and we greatly welcome the initiative of the Secretary of State for Energy in promoting wide public debate on this subject. In that debate and elsewhere([68]) it has been suggested that there is a need for a high-level independent body to provide advice on energy strategy, taking account of economic, social, technical and environmental considerations. We strongly support this idea. Our study of nuclear power has made us deeply aware of the need for authoritative exposition of the implications of different policies that might be adopted on energy supply. It is our particular concern that the environmental consequences of alternative strategies should be comprehensively explored and assessed, but this is only one aspect and the implications must be looked at as a whole. We do not think that this can be done adequately by existing machinery and by collective action of the government departments involved. There is a need for freedom and originality of thought, and for

independence from institutional pressures and from the need to justify existing arrangements and policies. It should be the principal aim of the advisory body to promote public understanding of the full implications of alternative energy strategies and to promote development of the consensus which is needed to guide the Government in framing its policies. We see a specific role for such a body in relation to proposals we make in paragraphs 523 and 524 regarding the future development of nuclear power.

FBR development

516. We have explained that we accept the need to maintain a nuclear option; our concern is that UK planning might depend too much on the introduction of FBRs on a substantial scale and a move into the plutonium economy. There is of course no present commitment to an FBR programme. The next major step in their development in this country and others is the design and construction of a first commercial scale fast reactor, the CFR1, which is needed in order to establish, among other things, whether the safety problems can be adequately resolved for commercial exploitation to be feasible. This development could well take a decade or more, and if it is accepted that the country may need FBRs in operation by about the turn of the century it is not too soon to start.

517. As a Commission concerned with environmental pollution it is not for us to propose that CFR1 should go ahead. Nor do we oppose the development of CFR1 itself on environmental grounds. Nevertheless, although it is a very long way from building CFR1 to a plutonium economy, we must view this highly significant first step with misgivings. The cost and momentum of technological development on this scale are such as to make one fear that it will lead inevitably to a large FBR programme in the future. There is a considerable risk that inadequate resources will be available for the development of alternative energy programmes, including conservation, which might have made substantial future dependence on fission power unnecessary.

518. The strategy that we should prefer to see adopted, purely on environmental grounds, is to delay the development of CFR1. This would provide time for the socio-political aspects of the plutonium economy to be fully investigated and debated, so that the risks could be widely understood and judged before this first step was taken. It would free resources and provide a stimulus for work on alternatives. It would give time for the problems of waste management and disposal to be properly investigated and, we hope, resolved. Each year of delay would increase the prospect of establishing whether a viable alternative strategy existed which would avoid the need for FBRs.

519. We recognise, however, that such a delay might well create very serious problems which are not within our remit for investigation. The prospect of indefinite postponement of a crucial stage in fast reactor development would inevitably result in the loss of expert technologists from the industry and weaken the sense of purpose which is essential to the enterprise. It might, therefore, seriously damage the prospects of eventual success. There is the possibility that delay might lose us opportunities for collaboration with other countries which

could reduce the cost of the development. There is the risk (notwithstanding our doubts about the future necessity for nuclear power on the scale of official projections) that the FBR will prove to be essential for vital energy supplies and that it will not be available in time. These are matters for the Government to consider in reaching a decision on CFR1.

520. We hope that in reaching that decision the Government will take account of our views as expressed in this Report, and particularly of the following:—

(i) It should be an aim of present policy to seek to lessen the possible future need for dependence on FBRs. Accordingly, the programme of research on other energy sources should be reappraised and should be pursued with more urgency than is now apparent and with a substantial increase in resources.

(ii) A small proportion of the resources that would be required for CFR1 could have a major impact if applied in other areas. Considerable savings might result if CFR development were pursued in collaboration with other nations, and if consideration was given to possibilities for phasing the project in ways that would not seriously affect its momentum.

(iii) If CFR1 development proceeds, it should be seen as being in the nature of an insurance policy, albeit an expensive one, justified by the importance of risks relating to the security of the nation's future energy supplies. It should be clear that the development implies no judgment on the inevitability or acceptability of a future FBR programme. These are matters which should be assessed by the procedure we describe below, and we would wish to see a commitment by the Government to this procedure before a decision to proceed with CFR1 was taken.

Public assessment of nuclear power

521. We have explained our reasons for thinking that nuclear development raises long-term issues of unusual range and difficulty which are political and ethical, as well as technical, in character. We regard the future implications of a plutonium economy as so serious that we should not wish to become committed to this course unless it is clear that the issues have been fully appreciated and weighed; in view of their nature we believe this can be assured only in the light of wide public understanding. We are perfectly clear that there has so far been very little official consideration of these matters. The view that was expressed by the Department of Energy in their evidence to us was that there were reasonable prospects that the safety and environmental problems posed by nuclear power could be satisfactorily overcome and that, if this proved not to be so, other forms of energy would have to be used, or consumption somehow reduced. We see this as a policy that could lead to recognition of the dangers when it would be too late to avoid them. More is needed here than bland, unsubstantiated official assurance that the environmental impact of nuclear power has been fully taken into account.

522. The question arises of how the necessary public understanding is to be brought about. Considered judgment requires the weighing of many factors; estimates of future energy needs in relation to projections of economic growth, and the environmental consequences of different energy strategies and their estimated economic and social costs. There is a need, we believe, openly and deliberately to weigh the risks and costs of embarking on a major nuclear programme against those of not doing so.

523. We have concluded that a special procedure is needed. This should follow, in general principles, that accepted for environmental impact statements now required for certain major projects in the USA. A comprehensive document setting out the issues and the evidence should be published first in draft. We envisage that much of the evidence would be prepared by the proponents of nuclear development, the AEA, but plainly other contributions would be required. The statement must not be confined to the effects of the first stage of development, but must follow through to the furthest point to which our current knowledge can attain. The social and economic, as well as the scientific, technological and environmental problems must be fully set out.

524. This publication should be followed by a further stage in which the comments made by interested agencies and individuals should be presented and evaluated. These presentations should ultimately receive independent assessment and we think that this would be an entirely appropriate function for the advisory body we have referred to in paragraph 515. No doubt the introduction of such a procedure would present many difficulties which we have been unable to examine closely, but we are convinced about the need for it. The ultimate aim is clear; it is to enable decisions on major questions of nuclear development to take place by explicit political process. The discussion of the possible environmental consequences of nuclear development which we offer in this Report will, we hope, contribute to this process but it cannot complete it.

CHAPTER XI

SUMMARY OF PRINCIPAL CONCLUSIONS AND RECOMMENDATIONS

525. In this chapter we list our principal conclusions and recommendations. The references given in brackets are to paragraph numbers.

526. **Radioactivity and radiobiology**

1. At the levels of radiation likely to be permitted in relation to possible somatic effects, the genetic effects should be of little concern (62).

2. No case has been made on the special carcinogenicity of "hot" particles of plutonium (69, 70). The current "maximum permissible body burden" for plutonium–239 is a reasonable level for skeletal deposition (72). A reduction by a factor of 5 of the "maximum permissible lung burden" for insoluble particles is already practised (77). On present evidence the derived standards for plutonium exposure and uptake are not seriously in error (77).

527. **Control arrangements**

3. There is no better way of deriving basic standards than on ICRP recommendations, given that the scientific standing and independence of its members is maintained (203).

4. The NRPB should have the statutory responsibility to advise the Government on basic standards proposed by ICRP and EURATOM (223). In view of the changes we propose in NRPB responsibilities, its composition and organisation should be reviewed (224) and the need for additional expertise within the DHSS should be considered by the Government (225).

5. The primary responsibility for assessing the effects of discharges of radioactivity on man and the environment should continue to rest with Government Departments. This responsibility should be independent of those of Government Departments deciding energy policy or licensing nuclear sites (241).

6. The present control arrangements have worked reasonably satisfactorily but changes are needed for the future (242, 243). A new Inspectorate (HMPI—as proposed in our Fifth Report) should be responsible for determining and controlling all discharges of radioactivity to the environment in consultation with MAFF and NRPB (247).

528. **Research**

7. There has been insufficient coordination of radiobiological research and we welcome the setting up by MRC and NRPB of a Joint Committee on Radiological Protection (228).

8. We were impressed with the quality of work on radioactivity in the marine environment by the FRL but it is important that the Government should encourage independent work in this field through the Research Councils (232, 233).

9. There has been insufficient research on radioactivity in the atmospheric and terrestial environments. The NERC should be funded to undertake this work in consultation with the ARC, the MRC and the NRPB (235, 236).

10. The NRPB should have responsibility for the broad programme of environmental research on radioactivity and for ensuring its adequacy having regard to expected developments in nuclear power. The NRPB should collaborate with the NERC and other appropriate bodies in identifying the research required. Research contracts should be open, with freedom to publish subject only to patent and security considerations (237).

11. There is a need for continuing research to assess the effects of radiation on the natural world (81, 238).

529. **Monitoring and surveys**

12. An annual survey should be published by HMPI of discharges of radioactivity to air, water and land (251). A comprehensive report on total radiation exposure should be published periodically by the NRPB (252, 253).

13. The NRPB should be responsible for initiating and coordinating monitoring in connection with environmental pathways (76, 252).

14. The national registry of radiation workers being established by the NRPB should cover ex-employees. Arrangements should also be instituted to monitor accidental exposures to radiation (253).

530. **Accidental releases of radioactivity**

15. The responsibility for specifying emergency reference levels should be vested in NRPB (254).

16. The emergency reference plan drawn up for a nuclear installation should be made available to interested members of the public who might be affected (255).

531. **Reactor safety and siting**

17. The risk of serious accident in any single reactor is extremely small; the hazards posed by reactor accidents are not unique in scale nor of such a kind as to suggest that nuclear power should be abandoned for this reason alone (176).

18. The criteria and methods of working of the NII should be reviewed by the authorities concerned (285–290).

19. There is a need for expert, independent advice to the Government on reactor safety and the HSE should develop the capability to provide this (292).

20. We should wish to see nuclear stations developed that could be sited in urban areas and used for district heating purposes (295).

21. There appear to be anomalies in current siting policy and the policy should be reviewed (296).

532. **Security and the safeguarding of plutonium**

22. Plutonium appears to offer unique potential for threat and blackmail against society because of its great radiotoxicity and its fissile properties (182, 322, 323).

23. The construction of a crude nuclear weapon by an illicit group is credible. We are not convinced that the Government has fully appreciated the implications of this possibility (325).

24. Given existing or planned security measures, the risks from illicit activities at the present level of nuclear development are small; the concern is with the future (308).

25. Plutonium extracted from fuel reprocessed for a foreign customer should, if returned, be incorporated in new fuel elements (319).

26. The unquantifiable effects of the security measures that might become necessary in a plutonium economy should be a major consideration in decisions on substantial nuclear development (332, 335). Security issues require wide public debate (336).

533. **Radioactive wastes**

27. There should be no commitment to a large programme of nuclear fission power until it has been demonstrated beyond reasonable doubt that a method exists to ensure the safe containment of long-lived, highly radioactive waste for the indefinite future (181, 338).

28. Radioactive emissions to atmosphere present no significant problems at present though a more systematic approach to control may be needed in the future (343–345). It will be necessary to take account of the possible additive effects of discharges; each nuclear site should have clear standards to which to work (346, 347).

29. A programme of research is required into the possible future effects of plutonium discharges to sea from Windscale (354).

30. The burial of low-level solid wastes at the Drigg site is satisfactory for the present but in the longer term such wastes should preferably be taken to a national disposal facility (360).

31. There is a lack of clearly formulated policy for the disposal of intermediate level solid waste at nuclear stations (364). The policy of accumulating more highly active solid wastes at AEA and BNFL sites with a view to eventual ocean disposal appears inadequate (367). Such disposal may prove unacceptable and the possible future requirements again point to the need for a national disposal facility (367, 372).

32. The probable benefits of actinide separations are uncertain and unlikely to justify delay in the vitrification programme. The issue requires consideration by an expert body (388).

33. It appears doubtful whether direct ocean disposal of vitrified high level wastes will be acceptable (396).

34. There are two reasonable options for the permanent disposal of vitrified wastes: to geological formations on land and below the ocean bed. But neither of these has been sufficiently studied nor demonstrated as a feasible option (397).

35. There should be a substantial UK effort in the field of disposal to geological formations on land and the required research should be carried out through the NERC and the IGS (406–408).

36. Disposal under the ocean bed has several advantages (411). A programme of research should be mounted through the NERC and the IOS (417, 418).

37. There are some issues concerned with the transport of radioactive materials which should be reviewed (421).

38. When foreign fuel is reprocessed in this country, environmental interests would not be served by the return of resulting wastes to the country of origin (423, 424).

39. The costs of sound waste management practices are unlikely to add appreciably to the costs of electricity generated from nuclear power (425, 426).

40. The responsibility for waste management strategy should rest with the Secretary of State for the Environment. A Nuclear Waste Management Advisory Committee should be established (428).

41. A Nuclear Waste Disposal Corporation should be established, charged with responsibility for the safe disposal of all waste arisings at nuclear sites (430–434).

534. **Energy strategy and the environment**

42. There are substantial environmental objections to a nuclear power programme on the scale envisaged in official projections (479–483).

43. It appears possible that an alternative strategy could be devised that would avoid the future need for a large nuclear programme based on fast reactors (492–497).

535. **Nuclear power and public policy**

44. The dangers of the creation of plutonium in large quantities in conditions of increasing world unrest are genuine and serious. We should not rely for energy supply on a process that produces such a hazardous substance as plutonium unless there is no reasonable alternative (506, 507).

45. The abandonment of nuclear fission power would, however, be neither wise nor justified (508). But a major commitment to fission power and a plutonium economy should be postponed as long as possible (511).

46. There should be increased support for the development of other energy sources including energy conservation, combined heat and power systems and fusion power (513).

47. The Government should consider the desirability of extending the functions of the AEA to cover work on renewable energy sources (514).

48. We support the need for an independent, high-level advisory body on energy strategy (515).

49. We do not oppose development of CFR1 itself on environmental grounds (517–519). However, our views should be taken into account by the Government in reaching a decision (520).

50. There should be no commitment to a large nuclear programme including fast reactors until the issues have been fully appreciated and weighed in the light of wide public understanding. A procedure for consultation is required to this end (521–524).

Acknowledgements

536. The many Government Departments, organisations and individuals we have consulted during our study are listed in Appendices 3 and 4. We are grateful to them all for the help and co-operation we have received. We must particularly express our appreciation to those bodies concerned with nuclear power in this country whose establishments we have visited. The time and trouble they have taken to help us to understand their operations have been remarkable and we could not have asked for more.

537. On some specialised aspects of our study we have appointed experts as consultants and we have greatly benefited from their advice. Those who assisted us in this way were Dame Janet Vaughan and Professor F. W. Spiers (on plutonium toxicity), Mr. F. R. Farmer (on reactor safety) and Professors P. J. Grant and W. B. Hall (on various aspects of nuclear technology). We wish to record our grateful thanks for their invaluable help. We are also indebted to Mr. David Gray of the Institute of Geological Sciences who accompanied some of us on our visit to the Asse Mine in West Germany and gave us much valuable help and advice.

538. It was announced in May 1976 that Professor Hans Kornberg had been appointed a member of the Commission and that he was to succeed Sir Brian Flowers as Chairman from September 1976. Professor Kornberg attended our last two meetings and although he had taken no part in our investigations he was present during our final discussions on the draft of our Report. We are very pleased to record that Professor Kornberg wishes to associate himself with our findings.

539. Finally, we wish to express our gratitude to all the staff of our small Secretariat. The demands we have placed on them during a long and complex study have frequently been severe but they have responded with unfailing willingness and efficiency. In particular, we must thank our Secretary Mr. Lionel Rutterford and our scientific assistant Dr. Grant Lewison; without their dedicated work on our behalf this Report could not have been prepared and we are keenly aware of our debt to them. Our study has involved us in many technical matters and we acknowledge a special debt to Dr. Lewison who has immersed himself deeply in the subject and on whose tireless researches we have greatly depended.

ALL OF WHICH WE HUMBLY SUBMIT FOR YOUR MAJESTY'S GRACIOUS CONSIDERATION

BRIAN FLOWERS (*Chairman*)
SHIRLEY ANGLESEY
EIRENE WHITE
RALPH VERNEY
RICHARD DOLL
FREDERICK WARNER
ERIC DENTON
DEREK BOWETT
TONY CHANDLER
JOHN COLLINGWOOD
TERENCE CONRAN
PATRICIA LINDOP
MURDOCH MITCHISON
RONALD NICOLL
RICHARD SOUTHWOOD
PAUL STREETEN

L. F. RUTTERFORD (*Secretary*)

D. J. MACVICAR (*Assistant Secretary*)

July 1976

One of our members, Mr. Frank Chapple, was unable to take a significant part in our deliberations because of his other commitments. In the circumstances he decided that he could not properly sign this Report.

APPENDIX 1

WRITTEN EVIDENCE SUBMITTED BY THE BRITISH INSTITUTE OF RADIOLOGY TO THE ROYAL COMMISSION ON ENVIRONMENTAL POLLUTION, MAY 1975

RADIATION EXPOSURE TO THE PUBLIC FROM MEDICAL RADIOLOGY

Whilst it must be realised that radiation exposure to the public from medical radiology should be considered separately from all other sources of radiation exposure, since medical radiography is undertaken *specifically to benefit the individual being exposed*, it is nonetheless proper to enquire whether in current radiological practice reductions can be made in radiation exposure to the population—and at what cost?

1. *What is the proportion of X-ray films which are spoilt and re-taken, and for what reasons?*

Members of the Radiation Protection Committee of the British Institute of Radiology have undertaken a survey during January and February 1975, of the numbers and causes of spoilt X-ray films in four X-ray departments which represent typical examples of the four major types of hospitals, *viz:* Teaching Hospital, District General Hospital (with School of Radiography), Accident and Emergency Service, and Private Nursing Home. The period during which *all* films taken in each department were reviewed was either one or two months. The data are summarised in Table I. In all four types of X-ray departments, the majority of films re-taken were required because of exposure faults—particularly in films taken with portable radiographic equipment. The next most common cause of failure was positioning faults, which included selection of the wrong beam-limiting cone or diaphragm setting, mis-positioning of the cassette relative to the X-ray beam, failure to push home the cassette tray, etc. The proportion of films which were rejected as being non-contributory to the examination was significant only in the teaching hospital in which a higher proportion of complex procedures were undertaken. Known persistent or recurrent machine faults in one department reflected the pressure of work which required continued use of a machine known to be troublesome, but machine faults did not make a large overall contribution to causing X-ray films to be re-taken. Patient movement and non-cooperation were of only relatively minor importance in causing X-ray films to be spoilt. The predominance of exposure faults makes it relevant to enquire into the costs of providing automatic exposure control in new X-ray apparatus; these are detailed in Annex 1. Whilst making similar provisions for at least some existing X-ray machines is possible, at *circa* £2,000 per machine it is clearly not economically feasible. It must also be pointed out that no suggestion is made for automatic exposure control for portable X-ray generators. These latter are a source of a high proportion of films spoilt in current practice, but portable X-ray generators are normally used only in the worst clinical situations with very ill patients whose prognosis may be very poor indeed. The costs of providing automatic limitation of X-ray field size and cassette position are also detailed in Annex 1. Such provisions are already made for all new undercouch tube systems used with image-intensifiers.

In summary, approximately one patient in seven who undergoes X-ray examination has at least one film spoilt and re-taken. Overall, however, this results in only *circa* 5–6 per cent increase in the population exposure to radiation which occurs as a result of medical radiography. Providing automatic controls of exposure and field size on all new X-ray machines, which could only reduce population exposure by less than *circa* 5 per cent of the dose resulting from medical radiography, would add, respectively *circa* 20–30 per cent and *circa* 15 per cent to their capital cost.

2. *Have there been significant changes in radiation dose per examination since the Adrian Committee Report, and what are the cost-benefit relationships for newer imaging methods?*

Since the Report of the Committee chaired by Lord Adrian in the late 1950s, the number of X-ray examinations carried out annually in the UK has increased by approximately 80 per cent. In 1960 radiographic work units in England and Wales were 21·78 million. In 1971 the figure was 37·15 million. This is an increase of 70·6 per cent and there has been a regular yearly increase since then. During the period 1960–71 the population increased by under 7 per cent, although with a somewhat larger proportion in the older age groups. This increase of X-ray examinations has not been confined to UK, in fact figures are higher for almost all other countries listed. In 1971 there were 335 examinations per thousand population in England and Wales, but between 400 and 600 per thousand in Belgium, Denmark, France, Finland, Holland and Sweden and 800–1,000 for Australia, Hungary and USA

The worst of the radiographic practices revealed by the Adrian Report have long since been rectified, and due to the appearance of new screens and X-ray films the average dose to the patient *per film* has decreased. However, in the absence of any firmly-based data, it is the impression of senior radiologists that the number of complex radiological examinations and hence the average number of films used per examination has increased. It would be surprising if the genetically significant dose to the population from medical radiography was decreased significantly today from the values reported in the Adrian Committee survey.

Ardran, Langmead and Crooks (1975)* have recently reported the dose reduction factors which might be possible in diagnostic radiography using

* Ardran, G. M., Langmead, W. A. and Crooks, H. E., 1975, Exposure reduction using new screen/film combination. *Brit. J. Radiol. 48;* 233-234.

newly-available rare earth oxysulfide screens in combination with green-sensitive X-ray films. It would appear that the application of these new screen-film combinations might reduce the population dose from medical radiography to about half the present level whilst maintaining an image quality adequate for all except very special situations (such as foetal radiography). Although the dose reduction per film would undoubtedly be greater than a factor two, this would be counter-balanced by the unchanged radiation doses required for fluoroscopic examination. The cost of a change to these new screen-film combinations would be approximately £110 per pair of screens. This would mean a cost of £5,000–£8,000 per average large department, or some £1·5 million taken over the whole country. This is obviously not practicable financially at this time, and the most that can be hoped is the limited use in average departments, leaving the majority of patients to receive larger doses from current film/screen combinations.

3. *What proportion of radiographic examinations is clinically avoidable?*

This question is largely unanswerable, as the number of divergent replies obtained will be in direct proportion to the number of doctors asked. In addition to differences in demands made for X-rays by examining physicians and surgeons which, in general, tend to decrease with increasing expertise on the part of the doctor, it must be recognised that (particularly in casualty situations) a significant number of clinically "unnecessary" radiographs are nonetheless required for medico-legal reasons to avoid later disputes over the extent of the injury sustained.

There is, in any case, some evidence to show that the number of films per examination can be reduced in certain circumstances. J. W. D. Bull, writing in the British Medical Journal of 0.11.68 (p.569), shows how radiological appraisal of the value of skull examinations in certain conditions enabled him to cut the number of views from four to one. He referred to this subject in the Langdon Browne lecture at the Royal College of Physicians (British Medical Journal, 10.8.74 p.394). A further reference from this lecture quotes a leader written by Dr. Stewart Mason in the Journal of the Royal College of Physicians. He questioned the value of radiological investigation of the renal tract in hypertension. The Royal College of Radiologists is at present collecting further information along these lines.

No quantitative estimates are available of the amount by which the use of diagnostic X-rays is being avoided because of the availability in the same hospital of alternative imaging techniques such as ultrasound (which exposes the patient to no ionizing radiation) or radioisotopes (for which the radiation doses to the patient are usually considerably lower than for radiographic examinations). These latter methods are, however, often complementary to radiographic examination rather than supplanting the need for X-rays.

One area in which concern is often expressed about the use of X-rays is in pregnancy, and members of the Institute's Radiation Protection Committee have undertaken during the past year a survey of the extent to which X-rays are being used for the assessment of foetal maturity, foetal abnormality and placental localization. Enquiries in centres throughout the country which are reasonably typical of radiological practice in both teaching and non-teaching hospitals show a situation which calls for thought. At eight centres which were questioned during 1973 and 1974, the percentage of deliveries in which abdominal X-rays were taken in late pregnancy varied from 8·6 to *35·3 per cent*, with an average of 22·7 per cent (see Table II). The senior clinicians in these units were under the impression that the figure was five to at most ten per cent! Because in recent years a higher proportion of patients is delivered in hospital, one would have expected that there would be a *lower* percentage of patients with potentially abnormal pregnancies who would be likely to require an abdominal radiograph. One worrying inference from the study is that where a patient attends a hospital with X-ray facilities easily available, these will be used—at Hospital 5 (Table II) the X-ray set was withdrawn and the patients sent to a nearby hospital; this small barrier to the use of X-rays may in part explain the low figure of 8·6 per cent of pregnancies radiographed at this centre. Certainly over the last 20 years, the number of X-ray films per obstetric examination has decreased from *circa* 3 to one, and, with concomitant increase in screen and film speeds, the radiation *dose per examination* has been reduced by a large amount. Thus the increase in the *number* of examinations may be more than off-set by these improvements of technique, but the need for radiography and the techniques used should be kept under review. As a general rule, no *non-obstetric* examination of the abdomen should be performed in pregnancy unless the patient's immediate or future life is threatened. Such occasions are very rare; only one having occurred in the last 15,000 deliveries at Hospital 8 (Table II).

It is clear that the younger generation of doctors needs education in the proper use of radiological facilities, *including* the possible hazards. This education must begin at undergraduate level as it is here that long-lasting impressions are gained. Such education requires properly constituted University Departments of Radiology. Parallel with the need for educating the radiologist is the similar need to educate the radiographer. Whatever guidelines may be laid down in any department, it is the radiographer who actually picks up the cassette and exposes the patient to X-rays. She, too, must be in full possession of the relevant facts and thereby appreciate the need to reduce foetal dose, or even (through the radiologist) question the need for a particular examination of the pregnant abdomen.

4. *Turning from radiodiagnosis to radiotherapy, has the marked improvement in the results of radiotherapy for Hodgkins Disease and related reticuloses increased significantly the population at risk of genetic damage from radiation exposure?*

An informal survey was carried out of the numbers of such patients treated during 1974 in radiotherapy departments in the UK from which members of staff are also members of the Institute's Committees. The consensus obtained pointed clearly to the relative rarity of this group of diseases, in comparison with the common cancers seen in their departments (e.g. carcinoma of bronchus, cervix, breast). It was felt that this very small population, while at an assessible level of personal risk of genetic damage, would not increase significantly the total pool of genetic abnormality in the general population. It must be pointed out that for the individual patient, the gain in being cured (in high percentage) of their disease far outweighs the risk of producing heritable genetic damage which might put their offspring at risk. Genetic counselling is already a regular feature of the treatment of such patients.

We are informed that a detailed study of patients with Hodgkins Disease treated by wide-field radiotherapy is being carried out by Miss F. E. Taylor, Environmental Studies Group, National Radiological Protection Board, Harwell, Didcot, Oxon.

This written evidence is submitted in response to specific questions raised by Professor P. J. Lindop, a member of the Royal Commission.

ANNEX 1

1. Automatic exposure control

Automatic exposure control in diagnostic X-ray equipment can be of the ionization chamber or photo-electric type. Most modern X-ray generators embody facilities such as a falling load system which makes the operation of automatic exposure control even more accurate. The cost of including such a system in new apparatus with, say, three working stations would be of the order of an additional 17–20 per cent of the cost of the total installation.

A further development of this system is generally termed "Automatic Programme Radiography", which sets out to eliminate the possibility of almost all operator errors. A number of push buttons are provided which are marked with the area of the patient to be X-rayed, e.g. wrist, thorax, etc. The pushing of the button selects the appropriate working site, the right tube focus, the previously agreed exposure factors, the setting of the density control, etc. Systems of this type have now been used successfully in many hospitals throughout the world, and in general add between 27 and 30 per cent to the cost of the complete installation.

2. Automatic collimation

All reasonably sophisticated fluoroscopic tables now embody automatic collimation, so that the beam-limiting diaphragm on the under-table tube is automatically adjusted to cover the image-intensifier field size selected, the cassette inserted into the serial film changer, the split-film programme selected, as well as adjusting for the distance between serial film changer and table top.

It is also feasible to add automatic collimators to over-table tubes, providing the equipment is already equipped with the necessary sensing devices to ensure alignment of tube and image receptor. The addition of such a system to a two working station, over-table X-ray installation would add approximately 12–15 per cent to its total cost.

TABLE I

Spoilt films in X-ray departments

Type of Hospital	*Number of Examinations Sampled*	*Average Number of Examinations per Patient*	*Number of Films Used*	*Average Number of Films per Examination*	*Number of Spoilt Films*	*Number of Patients with Spoilt Films*	*Major Causes of Spoilt Films*
1. Teaching	4,198	1·8	17,500	4·2	864(4·9%)	518(12·4%)	1. Exposure fault (47%) 2. Positioning fault (25%) 3. Film non-contributory to examination (11%)
2. District General	1,896	1·3	4,995	2·6	397(8·0%)	327(21·8%)	1. Exposure fault (33%) 2. Positioning fault (32%) 3. Machine fault (14%)**
3. Accident/Emergency	3,951	1·2	7,366	1·9	439(6·0%)	*(13·6%)	1. Exposure fault
4. Private Nursing Home	174	1·2	498	2·9	15(3·0%)	13(9·1%)	1. Exposure fault (87%) 2. Positioning fault (13%)

* Calculated value from average number of examinations per patient.
**Machine in one X-ray room was persistently faulty during period of survey.

TABLE II

Abdominal X-rays during pregnancy

(The figures from each hospital cover a period of twelve consecutive months; the starting date varied between 1st January 1973 and 1st January 1974)

Hospital No.	*No. of deliveries*	*No. of abdominal X-rays*	*% of deliveries X-rayed*
1	2,196	362	16·5
2	2,938	726	24·7
3	3,689	481	13·0
4	2,250	596	26·5
5	2,664	228	8·6
6	4,049	1,392	34·3
7	1,974	687	35·3
8	1,345	310	23·0
TOTAL	21,105	4,782	22·7

NOTES: (1) Ultra sound. Available to Nos. 2, 6.
(2) Isotopes. Available to Nos. 2, 3, 6, 7.
(3) Figures from two other Teaching Departments are known to be 18 per cent and 25 per cent but the actual number of deliveries and X-rays was not known.
(4) Hospitals Nos. 2, 5, 6, 7, are teaching units; Hospital No. 8 is a Central Regional Hospital; Hospitals Nos. 1, 3, 4, are peripheral Regional Hospitals.

APPENDIX 2

MEMBERS OF THE COMMISSION

This report was undertaken and completed by the Commission under the chairmanship of SIR BRIAN FLOWERS, MA, DSC, FINSTP, FRS, Rector of Imperial College of Science and Technology, and with the membership listed below. On 1 September 1976, having completed his term of office, Sir Brian Flowers was succeeded as Chairman by PROFESSOR HANS L. KORNBERG, MA, DSC, SC D, FIBIOL, FRS, Sir William Dunn Professor of Biochemistry, University of Cambridge and Fellow of Christ's College. The membership of the Commission was reconstituted from the same date.

THE MARCHIONESS OF ANGLESEY
Deputy Chairman of the Prince of Wales Committee
Chairman of the Welsh Arts Council

DR. D. W. BOWETT, MA, LLB, PHD, LLD
President of Queens' College, Cambridge

PROFESSOR T. J. CHANDLER, MSC, PHD
Professor of Geography, University of Manchester
Member of the Clean Air Council
Member of the Natural Environment Research Council

F. J. CHAPPLE, ESQ
General Secretary of the Electrical Electronic Telecommunication and Plumbing Union

DR. J. G. COLLINGWOOD, BSC, CENG, FI CHEM E
A director of Unilever
Fellow of University College London

T. O. CONRAN, ESQ
Designer
Chairman of Habitat

PROFESSOR E. J. DENTON, CBE, SC D, FRS
Secretary of the Marine Biological Association of the United Kingdom and Director of the Plymouth Laboratory
Honorary Professor, University of Bristol
Fellow of University College, London

PROFESSOR SIR RICHARD DOLL, OBE, DM, MD, DSC, FRCP, FRS
Regius Professor of Medicine, University of Oxford

PROFESSOR PATRICIA J. LINDOP, MB, PHD, DSC, MRCP
Professor of Radiation Biology, University of London, Medical College of St. Bartholomew's Hospital

PROFESSOR J. M. MITCHISON, SC D, FRSE, FI BIOL
Professor of Zoology, University of Edinburgh

PROFESSOR R. E. NICOLL, MSC, FRICS, FRTPI
Professor of Urban and Regional Planning, Strathclyde University

PROFESSOR T. R. E. SOUTHWOOD, PHD, DSC, ARCS, FIBIOL
Professor of Zoology and Applied Entomology, University of London
Chairman of the Division of Life Sciences and Director of Field Station, Imperial College of Science and Technology

P. P. STREETEN, ESQ, MA
Warden of Queen Elizabeth House
Director of the Institute of Commonwealth Studies
Fellow of Balliol College, Oxford

SIR RALPH VERNEY, Bt, KBE, JP
Forestry Commissioner
Chairman of the Secretary of State for the Environment's Advisory Committee on Aggregates

PROFESSOR SIR FREDERICK WARNER, DSC, CENG, FIMECHE, FICHEME, FRS
Senior Partner in Cremer and Warner (Consulting Engineers)
Visiting Professor in Chemical Engineering, Imperial College of Science and Technology
Visiting Professor in Environmental Engineering, University College London
Pro-Chancellor of the Open University
Chairman, British Standards Institution
Fellow of University College London

THE BARONESS WHITE, MA
Chairman of the Land Authority for Wales
Member of the British Waterways Board
Chairman of Advisory Committee on Oil Pollution of the Sea
Member of the Waste Management Advisory Council

APPENDIX 3

ORGANISATIONS AND INDIVIDUALS CONTRIBUTING TO THE STUDY

Those marked* gave oral evidence at meetings of the Commission, usually following a written submission.

A. Government and other organisations

*Advisory Council on Energy Conservation.

British Institute of Radiology.

*British Nuclear Fuels Ltd.

British Rail.

*Central Electricity Generating Board.

*Commission of the European Communities.

*Conservation Society.

Council for the Protection of Rural England.

Cumbria Area Health Authority.

*Department of Agriculture and Fisheries for Scotland.

*Department of Energy.

*Department of the Environment.

Department of the Environment for Northern Ireland.

Department of Industry.

Doctors and Overpopulation Group.

Foreign and Commonwealth Office.

*Friends of the Earth.

*Health and Safety Executive.

*H M Alkali and Clean Air Inspectorate, Health and Safety Executive.

*H M Industrial Pollution Inspectorate for Scotland, Scottish Development Department.

*Institute of Geological Sciences, Natural Environment Research Council.

*Institute of Oceanographic Sciences, Natural Environment Research Council.

Lake District Environmental Pollution Research Group.

Lancashire Area Health Authority.

*Medical Research Council.

Meteorological Office.

*Ministry of Agriculture, Fisheries and Food.

*National Coal Board.

*National Radiological Protection Board.

*Nuclear Power Company.

Royal Society for the Protection of Birds.

Scottish Development Department.

Society for Radiological Protection.

*South of Scotland Electricity Board.

The Radiochemical Centre.

*United Kingdom Atomic Energy Authority.

Welsh Office.

B. Individuals

Professor A. L. L. Baker, Emeritus Professor of Concrete Structures and Technology, Imperial College of Science and Technology, University of London.

Dr. V. T. Bowen, Woods Hole Oceanographic Institution, USA.

R. C. Burton, Esq.

Dr. Robert L. Drew.

George du Boulay, Esq, Consultant Radiologist, the National Hospital.

*Sir Kingsley Dunham, formerly Director of the Institute of Geological Sciences.

Dr. Michael Flood, Department of War Studies, King's College, University of London.

Henrik Harboe, Esq, formerly Managing Director, Stal-Laval (U.K.) Ltd.

Professor D. Leslie, Professor of Nuclear Engineering, Queen Mary College, University of London.

Dr. Peter Lindon.

Mrs. Elizabeth Marshall.

Sir Alec Merrison, Vice-Chancellor, Bristol University.

Sir Edward Pochin, Member of the National Radiological Protection Board and formerly of the Medical Research Council.

P. C. Roberts, Esq, Head of DOE Systems Analysis Research Unit.

*Dr. E. F. Schumacher, CBE, formerly adviser to the National Coal Board.

Mrs. A. Wilks.

John R. A. Young, Esq, Convenor, Transport and General Workers Union.

C. During the visits of the Commission to various localities, in addition to discussions with the staff of the organisations visited (listed in Appendix 4), views were put to the Commission by staff representatives and members of the local community. They were:

(i) Hinkley Point Nuclear Power Stations.

Members of the Local Liaison Committee.

Commander M. Ingham, Taunton Deane District Council.

D. J. Hunt, Esq, Chief Environmental Health Officer, Sedgemoor District Council.

R. J. Organ, Esq, Chief Environmental Health Officer, West Somerset District Council.

(ii) Windscale (BNFL)

Staff representatives:

J. Howard, Esq, Chairman, Windscale Branch, Institute of Professional Civil Servants.
W. Maxwell, Esq, Convenor, National Union of General and Municipal Workers.

Local representatives:

E. Bushby, Esq, farmer and National Farmers' Union representative.
Councillor J. J. Colligan, Mayor of Copeland Borough Council.
P. N. Denson, Esq, Chief Executive, Copeland Borough Council, (Member of Local Liaison Committee).
Dr. J. London, General Practitioner.
Superintendent Murphy, Cumbria Constabulary.
Rev. A. J. Postlethwaite, Vicar of Seascale.
Walter Thompson, Esq, Editor, "Whitehaven News".

APPENDIX 4

VISITS

Visits were made by groups of Commissioners to the following organisations:

Atomic Energy Research Establishment, Harwell.
Asse Salt Mine, West Germany.
Dounreay Experimental Reactor Establishment.
Hinkley Point Nuclear Power Stations.
MAFF Radiobiology Laboratory, Lowestoft.
National Radiological Protection Board.
Windscale (BNFL).
Winfrith Atomic Energy Establishment.

Visits were also made by the Secretariat to the following organisations:

Building Research Establishment, DOE.
Capenhurst (BNFL).
Dungeness Nuclear Power Station.
Energy Technical Support Unit, Harwell.
Institute of Geological Sciences
The Radiochemical Centre Ltd.
Springfields (BNFL).
Systems Analysis Research Unit, DOE.

GLOSSARY

1. Technical terms

actinides	Elements following actinium in the periodic table. They include uranium and plutonium. Many of them are long-lived α-emitters.
activation product	Material made radioactive as a result of irradiation, particularly by neutrons in a reactor.
binding energy	The energy theoretically needed to separate a nucleus into its constituent nucleons. It is a measure of stability of the nucleus.
blanket	Fuel elements in a fast reactor, that surround the core and contain depleted uranium a part of which is converted to the fissile plutonium–239 by the neutrons escaping from the core.
breed	To form fissile nuclei, usually as a result of neutron capture, possibly followed by radioactive decay.
breeding ratio	See conversion factor.
burn-up	Irradiation of nuclear fuel by neutrons in a reactor. It is measured in units of megawatt-days (of heat) per tonne of uranium or plutonium.
cave	A working space for the manipulation of highly radioactive items; it is surrounded by great thicknesses of concrete or other shielding and has deep protective windows.
cladding	Material used to cover nuclear fuel (uranium) in order to protect it and to contain the fission products formed during irradiation.
common-mode	Of failures, in which failure in one part of the system also affects the ability of another, supposedly independent part, to respond.
conversion factor	The number of fissile nuclei in irradiated fuel as a fraction of the number of fissile nuclei in new fuel.
core	The central region of a reactor where the nuclear chain reaction takes place, and heat is thereby generated.

curie	The unit of radioactivity in general use, corresponding to $3{\cdot}7 \times 10^{10}$ nuclear disintegrations per second, which is the amount of activity displayed by one gram of radium–226.
critical	Of an assembly of nuclear materials, that it is just capable of supporting a nuclear chain-reaction.
critical group	A group of members of the public who receive more radiation than any other people as a result of the discharge of a particular radioisotope to the environment at a site.
decay	Disintegration of a nucleus through the emission of radioactivity.
decay chain	A succession of radioisotopes, each one formed by decay of the previous one.
decay daughter	The disintegration product of a nucleus that has emitted radioactivity.
decay heat	Heat generated by the radioactivity of the fission products, which continues even after the chain reaction in a reactor has been stopped.
delayed	Of neutrons, that are emitted shortly after fission (the delay is of the order of tenths of a second).
depleted	Of uranium whose uranium–235 content is less than the 0·71% that occurs in nature.
dose commitment	Future radiation doses inevitably to be received, often because a particular radioisotope has been incorporated in body tissues.
elements	Parts of an assembly of nuclear fuel.
enrichment	The process of increasing the concentration of the uranium–235 isotope in uranium beyond 0·71% in order to make fuel made from it more suitable for use in a reactor.
environmental pathway	The route by which a radioisotope in the environment is transferred to man, e.g., by biological concentration in foodstuffs.

fast	Of neutrons, that they are travelling with a speed close to that with which they were ejected from the fissioning nucleus, typically 20,000 km/sec.
fast reactor	A reactor in which there is no moderator and in which the nuclear chain reaction is sustained by fast neutrons alone.
fault-tree	A diagram representing the possible initiating events and sequences of successive failures that would lead to a serious accident.
fertile	Of a nucleus, that it can become fissile by capture of one or more neutrons, possibly followed by radioactive decay; uranium–238 is an example.
fissile	Of a nucleus, that it will fission readily if it is struck by and captures a neutron.
fission	The splitting of a heavy nucleus into two (or more) parts, usually accompanied by a release of energy.
fission product	A nucleus of intermediate size formed from the breakdown or fission of a heavy nucleus such as that of uranium. Such a nucleus will be highly radioactive and usually emits β-particles.
fusion	The merging of two light nuclei to make a heavier one, usually with a release of energy.
genetic effects	Effects produced by radiation in the offspring of the person irradiated, usually malformations.
gonads	The organs (testes in men, ovaries in women) containing the reproductive cells (sperm, ova).
half-life	The time in which the number of nuclei of a particular type is reduced by radioactive decay to one-half.
heavy water	Water in which the hydrogen atoms all consist of deuterium, the heavy stable isotope which is present to the extent of 150 parts per million in ordinary hydrogen.
hex	Uranium hexafluoride, a corrosive gas (above 56 °C).
high-level	Of radioactive waste, that requires continuous active cooling in order to dissipate the internally-generated heat and prevent dissemination of the material.

hulls	Fuel elements or parts thereof from which the uranium has been dissolved out by acid, leaving only the cladding.
ion	An atom that has gained or lost one or more electrons and thus become electrically charged.
ionization	The creation of pairs of ions.
isomeric transition	The change from an excited state of an isotope to its ground state. It occurs with a characteristic rate, and is usually accompanied by the emission of γ-radiation.
isotopes	Two nuclei of the same chemical element that differ only in their mass. (Also used in place of "nuclides" in this report.)
light water	Ordinary water, used as moderator and coolant in some reactors called "LWRs".
maximum permissible body burden	The limit (usually the one recommended by ICRP) for a particular radioisotope in the body of a radiation worker that would cause him to be irradiated just to the level of the recommended basic standards.
maximum permissible concentration	The limit (normally the one recommended by the ICRP) for a particular radioisotope in air or drinking water, that for a worker exposed to such a concentration for 40 hours a week would cause him to be irradiated just to the level of the recommended basic standards.
metabolism	Behaviour within the body of chemical elements taken in by inhalation, ingestion etc.
mixed oxide	A mixture of plutonium and uranium dioxides, used as the fuel in fast reactors.
moderator	A substance used to slow down neutrons emitted during nuclear fission.
nuclear fuel cycle	The sequence of operations in which uranium is mined, fabricated into fuel, irradiated in a reactor, and reprocessed to yield uranium and plutonium for re-use as fuel.
nuclear material	Substances in which it is possible to induce some nuclear fissions, usually through the agency of a neutron or neutrons.

nuclear park	A large nuclear site in which there would be several large reactors, together with their associated fuel fabrication and reprocessing plants.
nuclide	Any particular type of nucleus, not necessarily radioactive, eg strontium–90.
nucleon	A neutron (uncharged) or a proton (positively charged).
plasma	A completely ionized gas at extremely high temperatures.
poisons	Substances that are strong absorbers of thermal neutrons and therefore make criticality less likely or reduce the amount of reactivity in fuel elements.
prefixes	tera (T) $\times 10^{12}$ pico (p) $\times 10^{-12}$ giga (G) $\times 10^{9}$ nano (n) $\times 10^{-9}$ mega (M) $\times 10^{6}$ micro (μ) $\times 10^{-6}$ kilo (k) $\times 10^{3}$ milli (m) $\times 10^{-3}$
prompt	Of neutrons, that are emitted immediately upon fission.
quality factor	A factor that attempts to account for the differing biological effectiveness of the various types of radiation. It is 1 for β- and γ-radiation, and 10 for α-radiation and neutrons.
rad	The unit of absorbed radiation, corresponding to 0·01 joules of energy per kg of material. (Radiation Absorbed Dose).
radioisotope radionuclide	A nucleus that is radioactive.
reactivity	Of fuel, its ability to support a nuclear chain reaction; it is a function of the concentration of fissile atoms and inversely of the quantity of neutron-absorbing material present.
rem	The unit of effective radiation absorbed by tissue; the product of the dose in rads and a quality factor (qv). (Röntgen Equivalent, Man; 1 röntgen=0·83 rad).
reprocessing	The chemical separation of irradiated nuclear fuel into uranium, plutonium, and radioactive waste (mainly fission products).

shielding	Material interposed between a source of radioactivity and an operator in order to reduce his radiation dose.
somatic effects	Effects produced by radiation in the body of the person irradiated, usually cancers.
spontaneous fission	The breaking of a heavy nucleus into two lighter ones without external initiation.
stringer	A mechanical link to join two fuel elements.
tailings	Crushed uranium ore from which the uranium has been extracted chemically.
tails	The depleted uranium produced at an enrichment plant, typically containing only 0·25 per cent of uranium–235.
thermal	Of neutrons, that they are travelling with a speed comparable with that of gas molecules at ordinary temperatures, about 2 km/sec.
vitrification	The incorporation of high-level wastes (mainly the oxides of metals formed as fission products) into glass.
weapons grade	Of uranium or plutonium, capable of being made into a nuclear assembly that would be critical on fast prompt neutrons alone.
yellowcake	A mixture of the two oxides of uranium, a yellow powder.
α-particle	A heavy, positively-charged particle; the nucleus of a helium–4 atom containing two protons and two neutrons.
β-particle	An electron; a light, negatively-charged particle.
γ-radiation	Electro-magnetic radiation of very short wavelength ($< 10^{-9}$m).
α-emitter	A radioisotope emitting α-particles.
β-emitter	A radioisotope emitting β-particles.
γ-emitter	A radioisotope emitting γ-radiation.

2. Acronyms and Abbreviations

ACAI	HM Alkali and Clean Air Inspectorate.
AEA	See UKAEA.
AEC	See USAEC.
AERE	Atomic Energy Research Establishment (Harwell).
AGR	Advanced Gas-cooled Reactor.
ARC	Agricultural Research Council.
AWRE	Atomic Weapons Research Establishment (Aldermaston).
BEIR	(Advisory Committee on the) Biological Effects of Ionizing Radiations.
BNDC	British Nuclear Design and Construction.
BNFL	British Nuclear Fuels Ltd.
bpm	best practicable means.
BWR	Boiling Water Reactor.
CANDU	CANadian Deuterium-moderated natural-Uranium fuelled reactor.
CEGB	Central Electricity Generating Board.
CFR	Commercial-scale Fast Reactor.
CHP	Combined Heat and Power.
Ci	Curie.
CRPPH	Committee on Radiation Protection and Public Health (EURATOM).
DAFS	Department of Agriculture and Fisheries for Scotland.
DE	Department of Employment.
DHSS	Department of Health and Social Security.
DOE	Department of the Environment.
DWL	Derived Working Limit or Level.
ECCS	Emergency Core Cooling Systems.
EEC	European Economic Community.
EIS	Environmental Impact Statement.
ERDA	(United States) Energy Research and Development Administration.
ERL	Emergency Reference Level (of radiation).
EURATOM	EURopean ATOMic Energy Community.
FAO	Food and Agriculture Organisation.
FBR	Fast Breeder Reactor.
FI	Factories Inspectorate.
FINGAL	~~Fission products INto GLAss.~~ Fixation in Glass of Active Liquor
FRL	Fisheries Radiobiological Laboratory (MAFF).

GNP	Gross National Product.
GSD	Genetically Significant Dose (of radiation).
GW	Gigawatts (10^6 kilowatts).
HARVEST	Highly Active Residues Vitrification Engineering STudies.
HMIPI	HM Industrial Pollution Inspectorate for Scotland.
HMPI	HM Pollution Inspectorate.
HSE	Health and Safety Executive.
HTR	High Temperature (gas-cooled) Reactor.
IAEA	International Atomic Energy Agency.
ICRP	International Commission on Radiological Protection.
IGS	Institute of Geological Sciences.
IOS	Institute of Oceanographic Sciences.
IPI	See HMIPI.
LMFBR	Liquid Metal-cooled Fast Breeder Reactor.
LWR	Light Water Reactor.
MAFF	Ministry of Agriculture, Fisheries and Food.
MBA	Materials Balance Area.
mca	maximum credible accident.
MPBB	Maximum Permissible Body Burden (of a radioisotope).
MPC	Maximum Permissible Concentration (of a radioisotope).
MPLB	Maximum Permissible Lung Burden (of a radioisotope).
MRC	Medical Research Council.
MUF	Material Unaccounted For.
MW	Megawatts (1000 kilowatts).
NAIR	National Arrangements for Incidents involving Radioactivity.
NEA	Nuclear Energy Agency (formerly European NEA).
NERC	Natural Environment Research Council.
NII	Nuclear Installations Inspectorate.
NNC	National Nuclear Corporation.
NPC	Nuclear Power Company Ltd.
NPT	(Nuclear) Non-Proliferation Treaty.
NRPB	National Radiological Protection Board.
NWDC	Nuclear Waste Disposal Corporation.
OECD	Organisation for Economic Cooperation and Development.
PFR	Prototype Fast Reactor (the 250 MW reactor at Dounreay).
PWR	Pressurised Water Reactor.
RCI	Radio-Chemical Inspectorate.

SDD	Scottish Development Department.
SGHWR	Steam Generating Heavy Water Reactor.
SI	Système Internationale.
SNG	Synthetic Natural Gas.
SRD	Safety and Reliability Directorate (AEA).
SSEB	South of Scotland Electricity Board.
TNPG	The Nuclear Power Group.
TRC	The Radiochemical Centre Ltd.
TWh	Terawatt-hour (10^9 kilowatt-hours).
UKAEA	United Kingdom Atomic Energy Authority.
UNSCEAR	United Nations Scientific Committee on the Effects of Atomic Radiation.
USAEC	United States Atomic Energy Commission.
WHO	World Health Organisation.

REFERENCES

1. OECD. *Energy Prospects to 1985*. OECD, Paris. 1974.
2. DEPARTMENT OF INDUSTRY. *Second Report on the Nuclear Ship Study*. HMSO, London. 1975.
3. POCHIN, E. E. *Estimated population exposure from nuclear power production and other radiation sources*. OECD, January 1976.
4. UNITED NATIONS SCIENTIFIC COMMITTEE ON THE EFFECTS OF ATOMIC RADIATION (UNSCEAR). *Ionizing radiation: levels and effects*. New York, 1972.
5. WEBB, G. A. M. *Radiation exposure of the public—the current levels in the United Kingdom*. National Radiological Protection Board publication R24, May 1974.
6. MITCHELL, N. T. *Radioactivity in surface and coastal waters of the British Isles, 1972-73*, MAFF Fisheries Radiobiological Laboratory report FRL 10, November 1975.
7. CLARKE, R. H. *An assessment of individual and collective doses due to Argon-41 discharges from CEGB Magnox reactors*. CEGB Berkeley Nuclear Laboratories report RD/B/N3483, September 1975.
8. KOCHUPILLAI, N. ET AL. *Down's syndrome and related abnormalities in an area of high background radiation in coastal Kerala*. Nature, vol. 262, 1 July 1976, pp. 60–61.
9. ADVISORY COMMITTEE ON THE BIOLOGICAL EFFECTS OF IONIZING RADIATIONS (BEIR). *The effects on populations of exposure to low levels of ionizing radiation*. US National Academy of Sciences and National Research Council, Washington, 1972.
10. TAMPLIN, A. R. AND COCHRAN, T. B. *Radiation standards for hot particles: a report on the inadequacy of existing radiation protection standards related to internal exposure of man to insoluble particles of plutonium and other alpha-emitting hot particles*. Petition to the AEC and EPA. Natural Resources Defense Council, Inc., Washington, 1974.
11. LOVINS, A. B. AND PATTERSON, W. C. *Plutonium particles: some like them hot*. Nature, vol. 254, 27 March 1975, pp. 278–80.
12. DOLPHIN, G. W. ET AL. *Radiological problems in the protection of persons exposed to plutonium*. NRPB publication R29, September 1974.
13. MEDICAL RESEARCH COUNCIL. *The toxicity of plutonium*. HMSO, 1975.
14. GOFMAN, J. W. *The cancer hazard from inhaled plutonium* and *Estimated production of human lung cancers by plutonium from worldwide fallout*. The Committee for Nuclear Responsibility, Dublin, California, May 1975. (Also printed in the US Senate Congressional Record, 31 July 1975, pp. 14610–20.)
15. MORGAN, K. Z. *Suggested reduction of permissible exposure to plutonium and other transuranic elements*. American Industrial Hygiene Association Journal, vol. 36, 1975, p. 565.
16. SPIERS, F. W. and VAUGHAN, J. *Hazards of plutonium with special reference to the skeleton*. Nature, vol. 259, 19 February 1976, pp. 531–4.
17. INTERNATIONAL COMMISSION ON RADIOLOGICAL PROTECTION. *Metabolism of plutonium and related elements and their compounds*. Report of a Task Group for Committee 2 of ICRP. Pergamon Press, Oxford, 1972.
18. VAUGHAN, J., BLEANEY, B. AND WILLIAMSON, M. *The uptake of plutonium in bone marrow—a possible leukaemic risk*. British Journal of Haematology, vol. 13, 1967, pp. 492–502.
19. TUCKER, A. *Figures show leukaemia link with plutonium workers*. The Guardian, 13 January 1975.
20. DOLPHIN, G. W. *A comparison of the observed and the expected cancers of the haematopoietic and lymphatic systems among workers at Windscale*. NRPB, September 1975, (to be published).
21. VAUGHAN, J. *Plutonium—a possible leukaemic risk*. To be published. 1976.
22. WILLRICH, MASON AND TAYLOR, THEODORE B. *Nuclear Theft: Risks and Safeguards*. Ballinger, 1974. See pp. 19–20.
23. TAYLOR, K. AND WALFORD, F. J. *Uranium from seawater: an energy cost study*. Programmes Analysis Unit, Chilton, Oxon, report R14/74, 1975.
24. CARRUTHERS, R. ET AL. *Fusion reactors and the environment*. Culham Laboratory Report CLM-M96, HMSO 1975.
25. AEC. *Proposed Final Environmental Statement for the Liquid Metal Fast Breeder Reactor Program* p.7.4.1. WASH-1535. December 1974.
26. WORKING PARTY OF THE COUNCIL FOR SCIENCE AND SOCIETY. *Superstar Technologies* Barry Rose (Publishers) Limited. 1976.
27. WEINBERG, A. M. *Social Institutions and Nuclear Energy*. Science. Washington, DC, vol. 177, no. 4043. 1972. p. 27.
28. ROSENBAUM, D. M. *A special safeguards study*. USAEC internal report to John F. O'Leary, Director of Licensing. Reprinted in Congressional Record (Senate), Vol. 120, no 59, 30 April 1974. pp. S 6621–30.

29. Royal Commission on Environmental Pollution. *First Report*. Cmnd. 4584. HMSO, London. February 1971.
30. Hafele and Sassin. *Energy Strategies* IIASA, Laxemburg. RR-76-8. March 1976.
31. Royal Commission on Environmental Pollution. *Air Pollution Control: an Integrated Approach*. Cmnd. 6371, HMSO, London. January 1976.
32. Health and Safety Executive. *The Explosion at Appleby-Frodingham Steelworks, Scunthorpe, 4 November 1975*. HMSO, London. 1976.
33. Porter, Sir George. Quoted in New Scientist, 24 April 1975, p. 189.
34. Nuclear Regulatory Commission. *Reactor Safety Study*. Wash–1400 1975. (The "Rasmussen" Report).
35. US Environmental Protection Agency. *Environmental radiation protection for nuclear power operations, proposed standards*. Federal Register, vol. 40, 29 May 1975, p. 104.
36. Kelly, G. N. et al. *The predicted radiation exposure of the population of the European Community resulting from discharges of Krypton–85, tritium, carbon–14 and iodine–129 from the nuclear power industry to the year 2000*. Commission of the European Communities. Doc. V/2676/75, Luxembourg, September 1975.
37. 111th Annual Report on Alkali, etc. Works, 1974. HMSO, 1975.
38. Hetherington, J. A. et al. *Environment and public health consequences of the controlled disposal of transuranic elements to the marine environment*. IAEA symposium on *Transuranic nuclides in the environment*, San Francisco, November 1975. IAEA-SM-199/11, pp. 139–54.
39. Bowen, V. T., Granada Television programme, *Nuclear waste in the Irish Sea*. 10 May 1976.
40. Bowen, V. T., Livingston, H. D. and Burke, J. C. *Distribution of transuranium nuclides in sediment and biota of the North Atlantic Ocean*. IAEA Symposium on *Transuranium nuclides in the environment*, San Francisco, November 1975. IAEA-SM-199/96, pp. 107–20.
41. Webb, G. A. M. and Morley, F. *A model for the evaluation of the deep ocean disposal of radioactive waste*. National Radiological Protection Board, report R14, June 1973.
42. Hetherington, J. A., Jefferies, D. F. and Lovett, M. B. *Some investigations into the behaviour of plutonium in the marine environment*. IAEA Symposium on *Impacts of nuclear releases into the aquatic environment*, IAEA-SM-198/29, pp. 193–212. See particularly the discussion on pp. 210–12.
43. Schulz, R. K., Tomkins, G. A. and Babcock, K. L. *Uptake of plutonium and americium by plants from soils*. IAEA Symposium on *Transuranium nuclides in the environment*, San Francisco, November 1975. IAEA-SM-199/92, pp. 303–10.
44. Blomeke, J. O., Nichols, J. P. and McClain, W. C. *Managing radioactive wastes*. Physics Today, August 1973, pp. 36–42.
45. Clelland, D. W. *Present methods of storing highly radioactive waste in the United Kingdom and proposals for the future*.
46. Claiborne, H. C. *Neutron-induced transmutation of high-level radioactive waste*. Oak Ridge National Laboratory—TM-3964, December 1972.
47. Gray, D. A. et al. *Disposal of highly active, solid radioactive wastes into geological formations—relevant geological criteria for the United Kingdom*. Inst. Geol. Sciences, report 76/12, HMSO, September 1976. (In press.)
48. Schneider, K. V. and Platt, A. M. (eds). *High Level Radioactive Waste Management Alternatives*. 4 vols. Battelle Pacific Northwest Laboratories, Richland, Washington, May 1974, report BNWL-1900.
49. Oswald, G. K. A. and Robin, G. de Q. *Lakes beneath the Antarctic ice sheet*. Nature, vol. 245, 5 October, 1973, pp. 251–4.
50. Ekren, E. B. et al. *Geologic and hydrologic considerations for various concepts of high-level radioactive waste disposal in coterminous United States*. US Geological Survey, open file report 74–158 (1974), 219 pp. Denver, Colorado.
51. Bishop, W. P. and Hollister, C. D. *Seabed disposal—where to look*. Nuclear Technology, vol. 24, December 1974, pp. 425–43.
52. USAEC Draft. *Generic Environmental Statement on Mixed Oxide fuel (GESMO)*. WASH–1327, August 1974.
53. USAEC Draft Environmental Statement on *Management of Commercial High-level and Transuranium-Contaminated Radioactive Wastes*. WASH-1539, September 1974 (Cost estimates on p. 9, 1–27).
54. Department of Energy. Energy Trends (monthly).
55. Building Research Establishment. *Energy conservation: a study of energy consumption in buildings and possible means of saving energy in housing*. BRE Current Paper 56/75, June 1975.
56. DOE, Central Unit on Environmental Pollution. *Effects of airborne sulphur compounds on forests and freshwaters*. Pollution Paper No. 7, HMSO, June 1976.

57. *BP Statistical review of the world oil industry 1975.* The British Petroleum Company Ltd., July 1976.
58. BROOKES, L. G. *The plain man's case for nuclear energy.* Atom, No. 234, April 1976, pp. 95–105.
59. USAEC. *Nuclear fuel supply.* WASH–1242, 1973.
60. AXTMANN, R. C. *Environmental impact of a geothermal power plant.* Science, vol. 187, No. 4179, 7 March 1975, pp. 795–803.
61. SHAW, T. L. (ED.). *An environmental appraisal of the Severn Barrage.* 1975.
62. DEPARTMENT OF ENERGY. *Energy R. & D. in the United Kingdom: a discussion document.* June 1976.
63. CHAPMAN, P. F. *Energy analysis of nuclear power stations.* Energy Policy, vol. 3 (No. 4) 1975, p. 285.
64. CHAPMAN, P. F. *Fuel's Paradise.* Penguin Books, 1975.
65. UK SECTION OF THE INTERNATIONAL SOLAR ENERGY SOCIETY. *Solar energy—a UK assessment.* The Royal Institution, May 1976.
66. SELECT COMMITTEE ON SCIENCE AND TECHNOLOGY REPORT. *Energy Conservation.* 11 September 1975. HMSO.
67. LOVINS, AMORY B. *World Energy Strategies: Facts, Issues, and Options.* Ballinger Publishing Company, Cambridge, Mass. 1975.
68. JOINT WORKING PARTY OF COENCO, ROYAL SOCIETY OF ARTS AND INSTITUTE OF FUEL. *Energy and the Environment.* Royal Society of Arts, London. 1974.

INDEX

(The references are to paragraph numbers. Footnotes are denoted by 'n')

Accidents, 10, 218, 254–7.
  Browns Ferry, 175, 310.
  Risk of, 168–76, 502.
  Windscale, 134, 300.
Actinides, 137–8, 303, 374.
  Separation of, 384–8.
Agricultural Research Council (ARC), 232, 234, 236, 428, 528.
Americium, 137–8, 303, 351, 353, 365, 371, 374–5, 384–7.
Argentina, 165 (Table 7).
Australia, 119, 208.
Austria, 165 (Table 7).

Bahamas, 121.
Belgium, 165 (Table 7), 372, 400.
Best practical means (bpm), 344, 345.
Bomb Tests—see Weapons, Nuclear.
Bradwell—see Central Electricity Generating Board.
Brazil, 165 (Table 7).
Breeding Ratio (conversion factor), 108–110.
British Institute of Radiology, 53, Appendix 1.
British Nuclear Design and Construction (BNDC), 279, 289.
British Nuclear Fuels Ltd (BNFL), 22, 73, 118, 129, 142, 237, 240, 250, 277, 331, 350–1, 354, 357, 364–5, 369, 372, 381, 383, 391, 407, 421–2, 428, 430–3, 462, 533.
  Works and Sites:
    Calder Hall, 103, 147, 277.
    Capenhurst, 20, 22, 118, 124, 357.
    Chapelcross, 103, 277.
    Drigg, 240, 357–60, 364, 369, 404, 533.
    Springfields, 20, 22, 118, 122, 126, 357.
    Ulnes Walton, 240, 357, 369.
    Windscale, 20, 233n, 277, 300, 309, 327n, 328–9, 373, 380.
      Discharges from, 80, 151, 250, 342, 350, 352–4, 368.
      Personnel, health of, 73–7.
      Plutonium fabrication, 22, 108, 118, 131–4, 144.
      Waste management, 129, 142, 365–8, 373, 380, 412, 533.
Bulgaria, 165 (Table 7).

Calder Hall—see British Nuclear Fuels Ltd.
Canada, 102 (Table 5), 106, 119, 165 (Table 7), 208, 400, 451.
Cancer, 50, 51, 52, 57, 67, 72, 77, 168, 322.
  Leukaemia, 51–3, 73–5, 170 (Table 8), 224, 230, 266.
  Lung cancer, 68–71, 120, 266, 303.
  Thyroid cancer, 266.
Capenhurst—see British Nuclear Fuels Ltd.
Chapelcross—see British Nuclear Fuels Ltd.
Central Electricity Generating Board (CEGB), 20, 237, 249, 279, 281, 289, 326, 349, 362, 364, 420, 431–2, 447, 480.
  Nuclear power stations:
    Bradwell, 281.
    Hartlepool, 279, 289.
    Heysham, 279.
    Hinkley Point, 279, 348.
    Oldbury, 103.
    Sizewell, 106.
    Trawsfynydd, 103 (Plate 1).
    Wylfa, 103, 128.

China, 450.

Combined Heat and Electrical Power Systems (CHP), 446, 465, 490–3, 513, 535.

Committee on Radiation Protection and Public Health (CRPPH), 208.

Curium, 137–8, 303, 374–5, 387, 395.

Czechoslovakia, 165 (Table 7).

Department of Agriculture and Fisheries for Scotland (DAFS), 240.

Department of Employment, 220 (Fig. 12).

Department of Energy, 21, 280, 291, 331, 422–3, 425, 428, 433, 466–9, 472, 477, 480, 491 493–4, 497, 509–15, 521.

Department of the Environment (DOE), 24, 219, 220 (Fig. 12), 234, 239–42, 246, 357, 420-1 428–32, 434–5, 533.

Department of Health and Social Security (DHSS), 24, 220 (Fig. 12), 223, 225–6, 527.

Department of Trade, 420.

Derived Working Limits (DWLs), 215–17, 241–5, 344, 345.

Didcot power station, 483.

Dounreay—see United Kingdom Atomic Energy Authority.

Drigg—see British Nuclear Fuels Ltd.

Economic growth, 8, 155, 477, 494.

Egypt, 165 (Table 7).

Emergency Reference Levels (ERLs), 218, 254, 530.

Energy, demand for, 8, 155–6, 450–4, 466–8.

Energy Research Development Administration (ERDA), 417, 426, 514.

Energy, sources of:
- Coal, 7, 85, 193–4, 438–9, 442, 447–9, 453, 460–1, Tables 12–13, 466, 472, 488-91
- Gas, 155, 438, 444, 448–9, 452–3, 460–2, Tables 12–13, 468–9, 472–3, 488, 490–1.
- Nuclear, 6–7, 155–6, 157–61, 190–5, 455–8, 462 (Table 13), 468–71.
- Oil, 7, 155, 194, 438, 442, 447–50, 452–3, 460–3, Tables 12–13, 469, 472, 488.
- Other, 156, 458–9, 486–7, 513–14.

Enrichment of uranium, 92.
- centrifuge process, 124.
- diffusion process, 123.
- laser process, 125.
- nozzle process, 125.

European Atomic Energy Community (EURATOM), 208–9, 214, 220, 329, 527.

European Commission, 210, 387, 391.

Factories Inspectorate (FI), 219, 220 (Fig. 12).

FINGAL process, 383.

Finland, 165 (Table 7).

Fisheries Radiobiological Laboratory (FRL), 220 (Fig. 12), 232–3, 236, 240, 245, 247, 251, 357, 528.

France, 102 (Table 5), 105, 112, 383, 400, 413, 491.

Fusion, 84, 145–52, 375, 469, 508, 511, 513, 535.

Genetic effects of radiation, 44, 57–65, 79.

Germany, West (Federal Republic of Germany), 107, 112, 165 (Table 7), 359, 360–1, 367, 400–1, 491, 513.

Hartlepool—see Central Electricity Generating Board.

HARVEST process, 383.

Harwell—see United Kingdom Atomic Energy Authority.

Health and Safety Commission, 280.

Health and Safety Executive (HSE), 21, 24, 219, 220 (Fig. 12), 278, 280, 290–2, 421, 531.

Heysham—see Central Electricity Generating Board.

Hinkley Point—see Central Electricity Generating Board.

HM Alkali and Clean Air Inspectorate (ACAI), 24, 219, 220 (Fig. 12), 240, 245–6, 344–5.

HM Industrial Pollution Inspectorate (IPI), 24, 220 (Fig. 12), 240.
HM Pollution Inspectorate (HMPI), 219, 246–7, 251, 343, 428, 527.
"Hot particles", 68, 69.
Hungary, 165 (Table 7).
Hunterston—see South of Scotland Electricity Board.

India, 59n, 165 (Table 7).
Institute of Geological Sciences (IGS), 359, 391, 397, 404–5, 407–8, 418, 533.
Institute of Oceanographic Sciences (IOS), 371–2, 396–7, 418, 533.
International Atomic Energy Agency (IAEA), 165, 206–8, 326–9, 371–2, 406, 420.
International Commission on Radiological Protection (ICRP), 23, 56, 67–8, 73, 78, 139, 187, 201–5, 207–11, 213–15, 220 (Fig. 12), 228, 244, 264, 344, 370–1, 527.
Iran, 165 (Table 7).
Irish Sea, 80, 139, 350, 352–4, 380.
Israel, 165 (Table 7).
Italy, 165 (Table 7), 400.

Japan, 5, 50, 59, 121, 165 (Table 7), 208, 409, 422, 423, 513.

Korea, 165 (Table 7).

Legislation (in chronological order):
 Emergency Powers Act 1920, 333.
 Radioactive Substances Act 1948, 23, 221.
 Atomic Energy Authority Act 1954, 20.
 Nuclear Installations (Licensing and Insurance) Act 1959, 21.
 Radioactive Substances Act 1960, 213–14, 239.
 Ionising Radiations (Sealed Sources) Regulations 1961, 214.
 Radiological Protection Act 1970, 23, 221, 223.
 Health and Safety at Work Act 1974, 214.
 Atomic Energy Authority (Special Constables) Act 1976, 334.
London Convention, 369–70, 380, 396.

Material Unaccounted For (MUF), 327.
Maximum Credible Accident (mca), 283–5, 296.
Medical Research Council (MRC), 23, 68–70, 73, 77, 220 (Fig. 12), 221, 223, 226, 228, 230, 236, 528.
Medical uses of radiation, 11, 24, 43, 45–6, 50, 53, Appendix 1.
Mexico, 165 (Table 7).
Ministry of Agriculture, Fisheries and Food (MAFF), 24, 80, 220 (Fig. 12), 231–2, 234, 239–40, 242, 245, 247, 343, 354, 357, 369, 372, 428, 432, 527.
Ministry of Defence, 22.
 Aldermaston, 22, 142.
Ministry of Supply, 20.

Namibia, 119.
National Arrangements for Incidents involving Radioactivity (NAIR), 256.
National Coal Board, 472, 488, 514.
National Nuclear Corporation (NNC), 279.
National Radiological Protection Board (NRPB), 23, 68, 70, 73–4, 76–7, 220 (Fig. 12), 221–30, 236, 237, 243, 247, 252–4, 256, 340, 354, 364, 370, 372, 391, 428, 432, 527–9.
National registry of radiation workers, 76, 253.
Natural Environment Research Council (NERC), 236–7, 354, 408, 428, 431–2, 528, 533.
Netherlands, 165 (Table 7), 372.
New Zealand, 208, 459.
Norway, 451.
Nuclear Energy Agency (NEA), 208, 369, 370, 372.
Nuclear Installations Inspectorate (NII), 21, 182, 219, 220 (Fig. 12), 253, 279, 280–2, 287, 290–1, 296, 364, 421, 531.

Nuclear Non-Proliferation Treaty (NPT), 165–7, Table 7, 326–9.
Nuclear parks, 177, 184, 312.
Nuclear Power Company (NPC), 21, 279, 289.
Nuclear priesthood, 184.
Nuclear Waste Disposal Corporation (NWDC), 430–5, 533.
Nuclear Waste Management Advisory Committee, 428–9, 432, 533.

Oldbury—see Central Electricity Generating Board.
Organisation for Economic Co-operation and Development (OECD), 208, 369, 450.

Pakistan, 165 (Table 7).
Philippines, 165 (Table 7).
Plutonium:
- Accidental release, 115, 303, 306.
- Danger from, 8, 25, 160, 316–25, 332, 384, 500, 505–7.
- Discharges to the environment, 350–4, 370–1.
- Diversion, 117, 182–6, 306, 316–25, 330–3.
- Fabrication plant, 143–4.
- Fissile properties, 95, 108, 132.
- Formation, 87, 108, 110, 113, 137, Table 6, 150, 457.
- Fuel for reactors, 110, 117, 130, 143–4, 159, 318, 319.
- Medical studies on, 54, 55, 229, 230.
- Other uses of, 131n.
- Radiotoxicity, 66–77, 143, 162, 182, 322.
- Safeguards for, 12, 135, 165–7, 183–5, 305–36, 506–7.
- Separation, 118, 130, 131n, 134–6, 166, 320.
- Stocks, 117, 135, 159–60, 311, 317, 411.
- Transport, 159, 160, 166, 183–4, 318–21, 423.
- Use in nuclear weapons, 91, 95, 130, 182, 322–5.
- Value, 182.
- Waste contaminated with, 136, 138, 144, 159, 178, 358, 364, 365–7.

'Plutonium economy', 161, 183, 185, 186, 306, 308, 321, 332, 496, 518.

Radioactive Substances Advisory Committee, 23.
Radioactive Waste Management Co-Ordinating Committee, 428.
Radiochemical Inspectorate (RCI), 24, 219 (Fig. 12), 239–40, 247, 356, 360, 428.
Radioisotopes:
- Americium-242, 35, 137 (Table 6).
- Argon-41, 340.
- Caesium-134, 351.
- Caesium-137, 80, 139, 264, 267, 303, 346n, 351, 370 (Table 10), 372, 384.
- Carbon-14, 341, 355.
- Cerium-144, 350.
- Cobalt-60, 362.
- Iodine-129, 56, 339.
- Iodine-131, 56, 128, 264, 266–7, 339, 379.
- Krypton-85, 98, 133, 302, 342.
- Neptunium-237, 131n, 137 (Table 6), 384.
- Neptunium-239, 87, 137 (Table 6).
- Niobium-94, 152.
- Plutonium-238, 66, 131n.
- Plutonium-239, 66, 72, 87, 95n, 108, 111n, 113, 131n, 137, 138, 150, 394.
- Plutonium-240, 95
- Potassium-40, 43 (Table 2)
- Radium-226, 33, 34, 35, 72.
- Radon-222, 34, 50, 120.
- Ruthenium-106, 80, 267, 350, 351.
- Strontium-90, 32, 46, 55, 78, 115, 350, 370 (Table 10), 372, 384.
- Thorium-230, 35.
- Thorium-232, 111.
- Thorium-233, 111.
- Thorium-234, 34.
- Tritium, 55, 78, 106, 145, 148–51, 341, 355, 366, 370 (Table 10), 371.

Radioisotopes—*continued*
Uranium-233, 111, 316.
Uranium-234, 123n, 137 (Table 6).
Uranium-235, 85–7, 90, 92, 94, 98, 104, 105, 108, 111n, 113, 123, 125, 130, 137 (Table 6).
Uranium-236, 130n, 137 (Table 6).
Uranium-238, 34, 87, 90, 94, 108–9, 111, 123, 125, 137, 145.
Uranium-239, 87.
Reactors, number of, 6, 158, 455–7.
Reactors, safety of, 8, 168–77, 260–304, 502–3.
Causes of accidents, 100, 101, 262, 309.
Comparison with other hazards, 170–3, 269–70, 275, 278, 290.
Consequences of accidents, 99, 168, 176, 263–8, 303–4, 310.
Design objectives, 173, 260, 271.
Doubts about, 101, 174–5, 276.
Fast reactors, 110, 112–17, 297–304.
Fault-free analysis, 272–6, 289.
Licensing process, 21, 279, 280–2, 287–8.
'maximum credible accident', 283–6, 296.
Siting policy, 293–6.
Thermal reactors, 105, 107, 301.
Reactor types:
AGR, 102 (Table 5), 104, 113, 123, 145, 195, 277, 279, 289, 293–4, 296, 301–2, 341, 345, 362–4.
BWR, 105, 106.
CANDU, 102 (Table 5), 106, 110–11.
CFR, 112–14, 128, 516–20, 535.
FBR, 22, 110, 112–17, 143, 292–3, 297–304, 312, 457, 502, 505, 510, 516–20.
HTR, 102 (Table 5), 107, 110–11.
LWR, 101, 102 (Table 5), 105, 131, 293n, 299, 301, 379.
Magnox, 102 (Table 5), 103, 104, 112–14, 118, 126, 128, 138, 277, 293–4, 301, 312, 340–1, 348, 362–3, 383, 426, 508.
PFR, 112, 114, 144, 275, 277, 304.
PWR, 105, 123.
SGHWR, 22, 102 (Table 5), 106, 114, 277, 477.

Safety and Reliability Directorate (SRD), 261, 277–8, 292, 296.
Scottish Development Department (SDD), 220 (Fig. 12), 240, 357, 359.
Security measures, see Plutonium, Safeguards for.
Select Committee on Science and Technology, 493.
Ships, nuclear propulsion, 11.
Siting of reactors, 158, 184, 255, 265, 293–6, 312, 362, 446, 478, 483, 503, 511.
Sizewell—see Central Electricity Generating Board.
South Africa, 165 (Table 7).
South of Scotland Electricity Board (SSEB), 20, 194, 279, 362, 432.
Nuclear power stations:
Hunterston, 279.
Torness, 106, 293.
Spain, 165 (Table 7).
Springfields—see British Nuclear Fuels Ltd.
Sweden, 165 (Table 7), 491.
Switzerland, 165 (Table 7).
Synthetic Natural Gas (SNG), 442, 447, 453, 476, 487n, 490–1.

Taiwan, 165 (Table 7).
Terrorism, 8, 160, 307, 323–5, 390, 505.
Sabotage, 10, 177, 306, 309–13, 336, 503.
Theft, 182–3, 306, 317–21, 326, 327, 332, 333, 336, 505.
Warfare, 177, 314–15, 503.
The Nuclear Power Group (TNPG), 21, 279.
The Radiochemical Centre Ltd. (TRC), 22.
Torness—see South of Scotland Electricity Board.
Trawsfynydd—see Central Electricity Generating Board.